Joint Varenna-Lausanne International Workshop on the Theory of Fusion Plasmas 2016

Journal of Physics: Conference Series Volume 775

Varenna, Italy
29 August – 2 September 2016

Editors:

Xavier Garbet
Olivier Sauter

ISBN: 978-1-5108-3388-3
ISSN: 1742-6588

Printed from e-media with permission by:

Curran Associates, Inc.
57 Morehouse Lane
Red Hook, NY 12571

CURRAN ASSOCIATES INC. proceedings.com

Some format issues inherent in the e-media version may also appear in this print version.

Printed by Curran Associates, Inc. (2017)

For permission requests, please contact the Institute of Physics
at the address below.

Institute of Physics
Dirac House, Temple Back
Bristol BS1 6BE UK

Phone: 44 1 17 929 7481
Fax: 44 1 17 920 0979

techtracking@iop.org

Additional copies of this publication are available from:

Curran Associates, Inc.
57 Morehouse Lane
Red Hook, NY 12571 USA
Phone: 845-758-0400
Fax: 845-758-2633
Email: curran@proceedings.com
Web: www.proceedings.com

Table of contents

Volume 775

Joint Varenna-Lausanne International Workshop on the Theory of Fusion Plasmas 2016

29 August to 2 September 2016, Varenna, Italy

Accepted papers received: 16 November 2016
Published online: 5 December 2016

Preface

Papers

Journal of Physics: Conference Series **775** (2016) 011001 doi:10.1088/1742-6596/775/1/011001

The joint Varenna-Lausanne international workshop on the theory of fusion plasmas took place in Varenna from August 29 to September 2 2016. Several issues of interest for fusion plasmas were addressed, namely MHD stability, RF heating, collisional and turbulent transport, plasma wall interaction, and physics of burning plasmas. The articles published in this special issue illustrate nicely the well balanced combination of physics, applied mathematics, and computer sciences that characterizes this workshop.

Let us mention several attractive topics, which are addressed in this issue. The question of 3D MHD equilibrium in tokamaks has received a great deal of attention, in connection with external resonant magnetic perturbations in tokamaks, and also stochastic edge in stellarators. The reader will also find some recent developments related to the effect of current drive and heating on the stability of tearing modes. As usual, turbulent transport is addressed in much detail. Several papers address specific numerical aspects of fluid and gyrokinetic codes, including code optimisation. Physics issues are abundantly dealt with, such as the impact of fast particles on turbulence, and particle transport. New numerical techniques to model wave propagation are presented, which provide significant advances in the field. Refinements such as the effect of density fluctuation on wave propagation, or the interaction between particles and the electromagnetic field near antennas, have also been studied in depth. Finally, specific issues such as nonlocal transport, decay of zonal flows, and the effect of neutrals on rotation have been investigated.

A striking feature of the 2016 edition was the large number of young faces among the participants. This is a great satisfaction for the organizers since a new generation of scientists is certainly needed whilst several devices come to operation, or will do so in a foreseeable future.

The diversity and quality of the papers published in the present cluster prove once more that the Varenna-Lausanne meeting has a special place among the conferences on magnetic fusion energy. It is certainly an excellent forum for exchanging ideas on theory and modelling, and for advertising the best results obtained in that field.

The present volume of Journal of Physics Conference Series contains 16 outstanding papers – all peer reviewed. Let us mention another set of about 20 papers to appear in Plasma Physics and Controlled Fusion. We hope the reader will appreciate this special issue and that it will be a source of inspiration for him.

Xavier Garbet, Olivier Sauter
November, 8 2016

Journal of Physics: Conference Series **775** (2016) 011002 doi:10.1088/1742-6596/775/1/011002

Peer review statement

All papers published in this volume of *Journal of Physics: Conference Series* have been peer reviewed through processes administered by the proceedings Editors. Reviews were conducted by expert referees to the professional and scientific standards expected of a proceedings journal published by IOP Publishing.

Kinetic modeling of 3D equilibria in a tokamak

C G Albert[1], **M F Heyn**[1], **S V Kasilov**[1,2], **W Kernbichler**[1], **A F Martitsch**[1], **A M Runov**[3]

[1] Fusion@ÖAW, Institut für Theoretische Physik – Computational Physics, Technische Universität Graz, Petersgasse 16, A-8010 Graz, Austria
[2] Institute of Plasma Physics, National Science Center "Kharkov Institute of Physics and Technology", Akademicheskaya Str. 1, 61108 Kharkov, Ukraine
[3] Max-Planck-Institut für Plasmaphysik, Wendelsteinstraße 1, 17491 Greifswald, Germany

E-mail: albert@alumni.tugraz.at

Abstract. External resonant magnetic perturbations (RMPs) can modify the magnetic topology in a tokamak. In this case the magnetic field cannot generally be described by ideal MHD equilibrium equations in the vicinity of resonant magnetic surfaces where parallel and perpendicular relaxation timescales are comparable. Usually, resistive MHD models are used to describe these règions. In the present work, a kinetic model is used for this purpose. Within this model, plasma response, current and charge density are computed with help of a Monte Carlo method, where guiding center orbit equations are solved using a semianalytical geometrical integrator. Besides its higher efficiency in comparison to usual integrators this method is not sensitive to noise in field quantities. The computed charges and currents are used to calculate the electromagnetic field with help of a finite element solver. A preconditioned iterative scheme is applied to search for a self-consistent solution. The discussed method is aimed at the nonlinear kinetic description of RMPs in experiments on Edge Localized Mode (ELM) mitigation by external perturbation coil systems without simplification of the device geometry.

1. Introduction

Resonant magnetic perturbations (RMPs) is presently one of the prospective tools for the control of edge localized mode (ELM) in a future tokamak-reactor [1]. Together with other external magnetic perturbations which violate the toroidal symmetry of a tokamak magnetic field such as toroidal field ripples and error fields, these perturbations may have a significant effect on plasma parameters, in particular, on plasma toroidal rotation velocity. It should be noted that, in most of the plasma volume away from resonant flux surfaces, RMPs can be well described by ideal MHD theory. In particular, if RMPs are strongly shielded by plasma response currents, ideal MHD computations using 3D equilibrium codes [2, 3] agree well with models that take non-ideal plasma response into account (e.g., resistive MHD models) in most of the plasma except for narrow resonant layers around resonant flux surfaces. This permits, in particular, to use externally perturbed ideal 3D MHD equilibria for computations of the non-resonant toroidal torque (neoclassical toroidal viscosity, NTV) resulting from the violation of the toroidal magnetic field symmetry [4]. At the same time, the resonant torque, which is comparable to NTV but, in contrast to NTV, is localized within resonant layers, is much more sensitive to the details of the electromagnetic field distribution in these layers and cannot be treated within ideal MHD. The same is true also for modification in the resonant layers of temperature and density profiles by RMPs that, together with the change in toroidal rotation (or, equivalently, of the equilibrium radial electric field), affects the RMP shielding by the plasma. The general trend of this modification is to bring plasma parameter profiles to the penetration point where the main shielding by electron component disappears (in the two fluid MHD theory this point corresponds to the zero of unperturbed perpendicular electron fluid velocity at the resonant flux surface [5, 6]), and, if RMP amplitude is above a certain threshold value, a bifurcation occurs [7] which results in the equilibrium state with nearly unshielded perturbation field and significant modification of the magnetic field topology (creation of islands). Clearly, the threshold RMP amplitude value predicted by the theory is sensitive to the accuracy of the non-ideal effects modeling, and often the very question whether RMPs are shielded or not in a particular experiment cannot be definitely answered taken the uncertainty of existing numerical models and the fact that typical RMP amplitudes are of the same magnitude order as the threshold amplitude values (see, e.g., Ref. [8]).

Presently the most advanced numerical models in terms of treatment of realistic 3D geometries and nonlinear effects are based on the MHD theory [9, 10, 11]. This theory, however, is not universally applicable to modern and, especially, future reactor scale devices where plasma collisionality may be very low. In this case MHD models using the realistic values of anomalous diffusion and shear viscosity often predict the linear width of resonant layers much below the ion Larmor radius in contrast to the kinetic model where these scales are comparable [12]. Besides that, collisionless effects in wave-particle interaction start to be important in resonant layers. A particular consequence of that is the shift of the field penetration point away from the zero of perpendicular electron fluid velocity, as predicted by the kinetic theory in presence of electron temperature gradient [8]. The magnitude of this shift depends on plasma collisionality. Finally, the diffusive ansatz in the description of anomalous transport employed in the MHD approach fails if the resonant layer width becomes comparable or smaller than the perpendicular correlation length of the turbulence (and that is not an exceptional case). Therefore, development of the 3D kinetic model of RMP interaction with the plasma which can address the above issues remains to be of interest.

An effort in this direction has been undertaken in Ref. [13] where parallel plasma current in a given perturbed field has been computed using a full-f guiding center PIC code, this current has been used then to compute the updated perturbed field (poloidal magnetic flux) and the convergence of this iterative procedure has been enabled by using the damped iteration scheme. The approach of the present report is based on the same idea of using the Monte Carlo method for evaluation of plasma response current combined with iterative account of plasma response field generated by this current. The main difference from Ref. [13] consists in using the δf method, which is better suited to weakly perturbed magnetic fields where it strongly reduces the requirements to the statistics as compared to the full-f method, and a preconditioned iteration scheme, which enables the convergence in cases where direct iterations are strongly linearly divergent, which is actually the case here. For those reasons the approach presented here is expected to be more efficient in terms of CPU cost.

The main purpose of the present report is to demonstrate the feasibility of such an approach and its relatively low CPU cost, therefore, the physical model has been strongly simplified by restricting the consideration only to the linear case (infinitesimal perturbations) and by treating the perturbation field in the leading order over the Larmor radius. In addition, modification of electrostatic field by the perturbations has been ignored here for simplicity. These and other problem simplifications do not principally affect the linear convergence of iterations and the CPU cost, which are the main points of interest in this initial study.

Journal of Physics: Conference Series **775** (2016) 012001 doi:10.1088/1742-6596/775/1/012001

2. Theoretical background

Presenting the total tokamak magnetic field as $\mathbf{B} = \mathbf{B}_0 + \delta\mathbf{B}$ where $\mathbf{B}_0$ is the axisymmetric equilibrium magnetic field and $\delta\mathbf{B}$ is the non-axisymmetric perturbation field fixed by the condition that the toroidal average of its components in cylindrical coordinates, $x^k = (R, \varphi, Z)$, associated with the main torus axis is zero, the later field is convenient to split into the known vacuum field $\delta\mathbf{B}_v$ and the unknown plasma response field $\delta\mathbf{B}_p$. Presenting the non-axisymmetric part of the plasma current as $\delta\mathbf{j} = \hat{P}\delta\mathbf{B}$ where $\hat{P}$ is the (linear in our example) integral operator corresponding to the solution of the kinetic equation, and presenting the solution to Maxwell equations in a similar form, $\delta\mathbf{B}_p = \hat{M}\delta\mathbf{j}$, the integral equation for the plasma response field, $\delta\mathbf{B}_p$, is

$$\delta\mathbf{B}_p = \hat{K}\delta\mathbf{B}_p + \delta\mathbf{B}_{pv}, \tag{1}$$

where $\hat{K} = \hat{M}\hat{P}$ and $\delta\mathbf{B}_{pv} = \hat{K}\delta\mathbf{B}_v$. Eq. (1) formally defines direct iterations, which are divergent if the spectrum $\hat{K}$ contains eigenvalues of modulus larger than one. As this is the case in the present problem, preconditioned iterations,

$$\delta\mathbf{B}_p^{(k+1)} = \delta\mathbf{B}_p^{(k)} - \hat{\Pi}^{-1}\left((\hat{K} - \hat{I})\delta\mathbf{B}_p^{(k)} + \delta\mathbf{B}_{pv}\right), \tag{2}$$

are used instead, where $\hat{I}$ is unity and the preconditioner $\hat{\Pi}$ is constructed from an approximation of the unstable part of the spectrum of $\hat{K}$ by a Krylov method (see, e.g., Ref. [14]). This method follows a similar approach as described in Ref. [15]. In addition to stabilization, it can also provide convergence acceleration and is used for this purpose in the NEO-2 code [4].

2.1. Non-axisymmetric plasma response current

In case of small enough perturbations and negligible finite Larmor radius effects (which is well fulfilled for electrons mainly responsible for the RMP shielding) the gyrokinetic equation linearized with respect to the perturbation field is

$$\frac{\partial\delta f}{\partial t} + V_0^k\frac{\partial\delta f}{\partial z^k} - \hat{L}_C\delta f = -\delta V^k\frac{\partial f_0}{\partial z^k}, \tag{3}$$

where $\delta f = f - f_0$ is the perturbation of axisymmetric steady state distribution function f_0, $\delta V^k = V^k - V_0^k$ is the non-axisymmetric perturbation of the phase space velocity and $\hat{L}_C$ is the linearized collision operator (in this case a Fokker-Planck operator described in Ref. [16]). Using for f_0 a local Maxwellian, $f_0 = f_M(\psi, W)$ where ψ is the unperturbed poloidal magnetic flux and W is the total particle energy (such an f_0 model approximates the core plasma but is not suited for the scrape-off layer and private flux region where, in turn, plasma response currents are relatively small), and taking δV^k in leading order over the Larmor radius (see, e.g., Ref. [8]), the r.h.s. of (3) is approximated as

$$-\delta V^k\frac{\partial f_0}{\partial z^k} \approx \dot{w}_s f_M, \qquad \dot{w}_s \equiv -v_\parallel\frac{\delta B^\psi}{B_0}\left(\mathcal{A}_1 + \mathcal{A}_2\frac{m_\alpha v^2}{2T_\alpha}\right), \tag{4}$$

where v and $v_\parallel$ are velocity module and parallel component, respectively, $\delta B^\psi = \delta\mathbf{B}\cdot\nabla\psi$ is the normal perturbation magnetic field component, and the thermodynamic forces are

$$\mathcal{A}_1 = \frac{\partial\ln n_\alpha}{\partial\psi} + \frac{e_\alpha}{T_\alpha}\frac{\partial\Phi_0}{\partial\psi} - \frac{3}{2}\frac{\partial\ln T_\alpha}{\partial\psi}, \qquad \mathcal{A}_2 = \frac{\partial\ln T_\alpha}{\partial\psi}. \tag{5}$$

Here, Φ_0 is the unperturbed electrostatic potential, and e_α, m_α, n_α and T_α are α species charge, mass, density and temperature, respectively. Within a δf Monte Carlo algorithm, the solution to (3), which determines the non-axisymmetric plasma response currents, is given by the expectation value of the test particle distribution function,

$$\delta f(\mathbf{z}) \approx \frac{1}{t_0}\int_0^{t_0} \mathrm{d}t\, \overline{w\,\delta\left(\mathbf{z}-\mathbf{Z}(t,\mathbf{z}_0)\right)} \tag{6}$$

where $\mathbf{Z}(t,\mathbf{z}_0)$ is the stochastic orbit of the unperturbed motion corresponding to the l.h.s. of (3) and originating at the phase space point $\mathbf{z}_0$ being also the random number distributed in accordance with f_M, test particle weight w is evolving in time in accordance with $\dot{w} = \dot{w}_s(\mathbf{Z}(t,\mathbf{z}_0))$, and particle tracing time t_0 is chosen to be much longer than particle correlation time with the perturbation field.

2.2. Non-axisymmetric part of the magnetic field

For the computation of the nonaxisymmetric part of the magnetic field in cylindrical coordinates $x^k = (R,\varphi,Z)$, it is convenient to use a Fourier expansion of the covariant components of the vector potential in φ with $\tilde{A}_k = \sum_{n\neq 0} A_{k,n} e^{in\varphi}$. The gauge $\tilde{A}_\varphi = 0$ is chosen as described e.g. in the appendix of Ref. [12], leading to a particularly simple representation of Fourier amplitudes of the contra-variant components (equal to the physical ones in R and Z) of the non-axisymmetric part of the magnetic field perturbation,

$$\delta B_n^R = \frac{in}{R} A_{Z,n}, \qquad \delta B_n^\varphi = (\nabla\times\tilde{\mathbf{A}})_n^\varphi = \frac{1}{R}\left(\frac{\partial A_{R,n}}{\partial Z} - \frac{\partial A_{Z,n}}{\partial R}\right), \qquad \delta B_n^Z = -\frac{in}{R} A_{R,n}. \tag{7}$$

From Ampère's law, a 2-dimensional system of differential equations in $A_{R,n}, A_{Z,n}$ follows as

$$-\frac{\partial}{\partial Z}\left(R^2(\nabla\times\tilde{\mathbf{A}})_n^\varphi\right) + \frac{n^2}{R}A_{R,n} = \frac{4\pi}{c}Rj_n^R, \qquad \frac{\partial}{\partial R}\left(R^2(\nabla\times\tilde{\mathbf{A}})_n^\varphi\right) + \frac{n^2}{R}A_{Z,n} = \frac{4\pi}{c}Rj_n^Z, \tag{8}$$

where j_n^R and j_n^Z are Fourier amplitudes of current density components. The equivalent weak form for the homogenous Dirichlet problem ($\tilde{\mathbf{A}}$ orthogonal to the boundary of the computation domain Ω) of Eqs. (8) is

$$\begin{aligned}\int_\Omega \mathrm{d}R\,\mathrm{d}Z R\left(R^2(\nabla\times\tilde{\mathbf{A}})_n^\varphi(\nabla\times\mathbf{W})^\varphi + \frac{n^2}{R^2}\left(W_R A_{R,n} + W_Z A_{Z,n}\right)\right) \\ = \frac{4\pi}{c}\int_\Omega \mathrm{d}R\,\mathrm{d}Z\, R\left(W_R j_n^R + W_Z j_n^Z\right),\end{aligned} \tag{9}$$

where a test function $\mathbf{W}$ with $W_\varphi = 0$ is chosen from the appropriate function space. If Ω is chosen with a large enough current-free region around the actual domain of interest, this description is suited to approximately describe the decay of the magnetic field at infinite distance.

3. Numerical implementation and results

The Monte Carlo procedure for evaluation of the perturbed distribution (6) and corresponding plasma response currents has been realized on the basis of the K2D code [16] which employs a geometrical integrator [17] for advancing the guiding center orbits with a significantly lower CPU cost as compared to usual integration of guiding center equations. Various quantities are discretized in this code on the triangular mesh (left Fig. 1) covering the whole vacuum vessel volume including the divertor region and the scrape-off layer. Eq. (9) is solved by a Galerkin finite element method using 2-dimensional triangular edge elements [18] of lowest order within the package *FreeFEM++* [19]. The Dirichlet problem is solved on an extended mesh (right Fig. 1) with a vacuum region around the original domain to emulate the current-free space in the

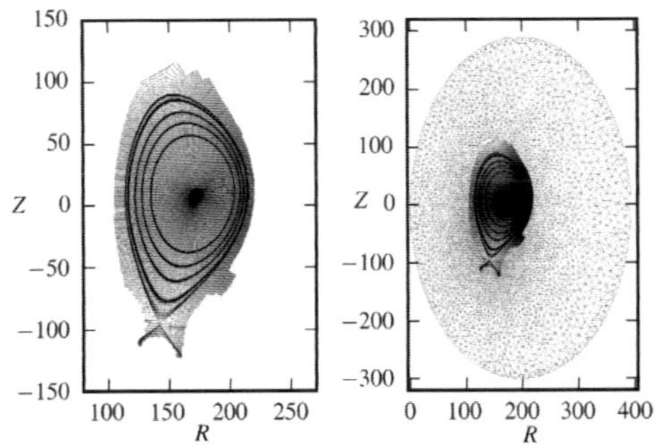

Figure 1. Mesh for the Monte Carlo code (left) with grid refinement near resonant flux surfaces and extended Mesh for the field solver (right).

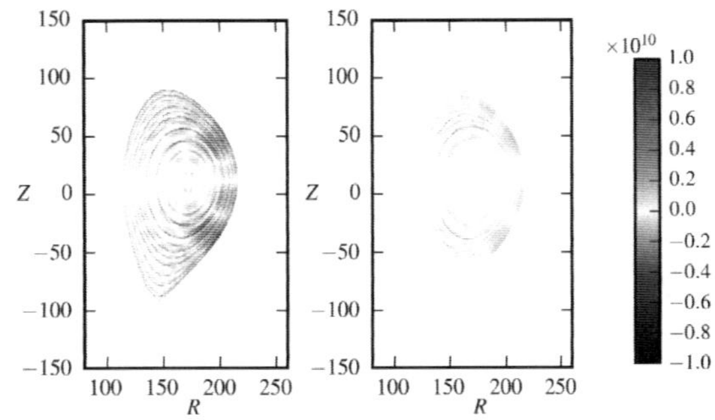

Figure 2. Real part of Fourier amplitude of cylindrical radial component of response current density j_n^R for electrons (left) and ions (right).

exterior of the device. While this region is not large enough for a highly accurate representation of vacuum boundary conditions here, it is sufficient for avoiding a strong influence from the domain boundary for the intended proof of concept.

Here, we consider a 3D equilibrium for a case corresponding to a typical medium size tokamak configuration with ELM mitigation coils producing a resonant magnetic perturbation of toroidal mode number n=2. Within the linear approximation described by Eq. (3), perturbation fields and corresponding plasma response currents are treated as infinitesimals. Therefore, coupling of toroidal Fourier modes (nonlinear effect) is negligible, and the solution for the currents and fields is fully described by Fourier amplitudes of the toroidal mode $n = 2$ shown in Figs. 2-5. Quasilinear effects of small but finite perturbation field on plasma parameter profiles (see, e.g., Ref. [8]) have been out of interest in the present initial study since they do not affect the linear convergence. Therefore, profiles of temperature, density and electric potential are modeled as linear functions of the poloidal magnetic flux ψ with magnitudes consistent with plasma parameters of present day devices.

The resulting plasma response currents are strongly localized around rational surfaces, which is illustrated in Fig. 2. The currently used mesh is not yet fully capable of resolving all details in the radial direction. However, the results for the magnetic field perturbation (Figs. 3-5) show the typical behavior observed in earlier studies by MHD calculations [20, 21]. While the plasma response leads to strong reduction of the perturbation fields around rational surfaces that are resonant with the perturbation field, amplification is observed outside this region (Figs. 4 and 5). This is in qualitative agreement to the results in Ref. [21].

4. Conclusion and outlook

In this article, the computation of a linear kinetic three-dimensional equilibrium in a tokamak with resonant non-axisymmetrical magnetic field perturbations has been demonstrated. The results were obtained by a kinetic δf Monte Carlo code using a geometric orbit integrator in conjunction with a finite element field solver employing two-dimensional edge elements for the computation of non-axisymmetrical toroidal perturbation harmonics. In the iterative scheme resulting from coupling these two models, convergence could be achieved by a preconditioner based on a Krylov subspace method.

The obtained results show typical shielding near resonant rational surfaces and are in qualitative agreement with existing results. Potential features of the presented method that allow to overcome some of the limitations of MHD models, especially in narrow resonant layers

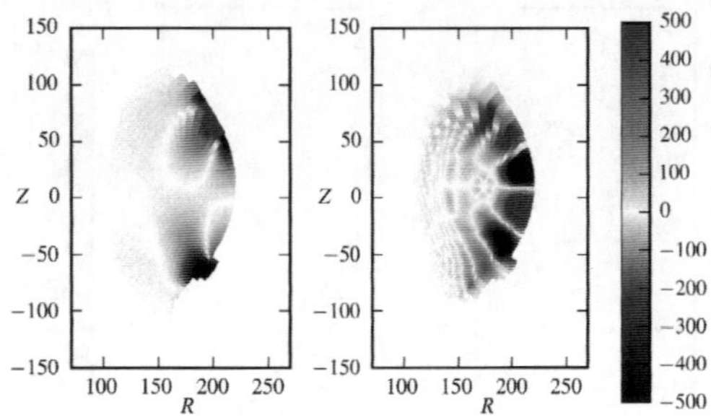

Figure 3. Real part of contra-variant radial component of magnetic field perturbation δB_n^ψ for the vacuum field (left) and for the field including the plasma response (right).

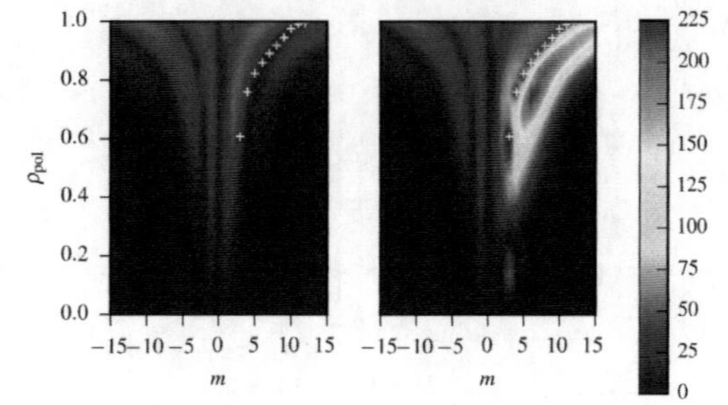

Figure 4. Modules of poloidal harmonics of δB_n^ψ in symmetry flux variables for vacuum field (left) and including plasma response (right) depending on $\rho_{\rm pol}$ (see Fig. 5). "+" marks resonant surfaces $m + nq = 0$ where shielding occurs.

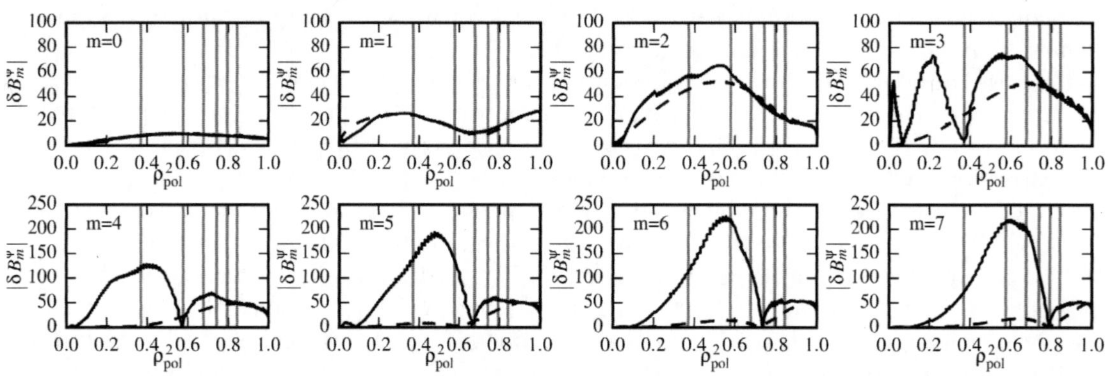

Figure 5. Module of poloidal modes of the radial contra-variant magnetic field component δB_n^ψ including plasma response (solid) and vacuum field (dashed) over $\rho_{\rm pol}^2 = (\psi - \psi_{\rm axis})/(\psi_{\rm sep} - \psi_{\rm axis})$. Rational surfaces with $q = 1.5, 2, 2.5, 3, 3.5, 4$ are marked by vertical gray lines. Shielding by plasma currents leads to vanishing perturbation fields at radial positions where $m + nq = 0$.

around rational surfaces, have been pointed out. Since the models do not rely on magnetic coordinates, the approach is well suited for investigations of three-dimensional effects in the region of the divertor and the scrape-off layer.

To obtain quantitative results for comparison to experiments, non-linear modeling will be required. This method will use a non-linear δf algorithm combined with a 3D geometric integrator.

Acknowledgments
This work has been carried out within the framework of the EUROfusion Consortium and has received funding from the Euratom research and training programme 2014-2018 under grant agreement No 633053. The views and opinions expressed herein do not necessarily reflect those of the European Commission. The authors gratefully acknowledge support from NAWI Graz and funding from the OeAD under the grant agreement *Wissenschaftlich-Technische Zusammenarbeit mit der Ukraine* No UA 06/2015.

References

[1] Evans T E, Moyer R A, Burrell K H, Fenstermacher M E, Joseph I, Leonard A W, Osborne T H, Porter G D, Schaffer M J, Snyder P B *et al.* 2006 *Nat. Phys.* **2** 419–423
[2] Hirshman S P, van RIJ W I and Merkel P 1986 *Comp. Phys. Commun.* **43** 143 – 155
[3] Strumberger E, Günter S and Tichmann C 2014 *Nucl. Fusion* **54** 064019
[4] Martitsch A F, Kasilov S V, Kernbichler W, Kapper G, C G Albert C G, Heyn M F, Smith H M, Strumberger E, Fietz S, Suttrop W, Landreman M, the ASDEX Upgrade Team and the EUROfusion MST1 Team 2016 *Plasma Phys. Contr. Fusion* **58** 074007
[5] Waelbroeck F L 2003 *Phys. Plasmas* **10** 4040–4047
[6] Waelbroeck F L, Joseph I, Nardon E, Bécoulet M and Fitzpatrick R 2012 *Nucl. Fusion* **52** 074004
[7] Fitzpatrick R 1998 *Phys. Plasmas* **5** 3325–3341
[8] Heyn M F, Ivanov I B, Kasilov S V, Kernbichler W, Leitner P, Nemov V V, Suttrop W and ASDEX Upgrade Team 2014 *Nucl. Fusion* **54** 064005
[9] Nardon E, Tamain P, Bécoulet M, Huysmans G and Waelbroeck F 2010 *Nucl. Fusion* **50** 034002
[10] Izzo V and Joseph I 2008 *Nucl. Fusion* **48** 115004
[11] Liu Y, Kirk A, Sun Y, Cahyna P, Chapman I, Denner P, Fishpool G, Garofalo A, Harrison J, Nardon E *et al.* 2012 *Plasma Phys. Contr. Fusion* **54** 124013
[12] Heyn M F, Ivanov I B, Kasilov S V, Kernbichler W, Joseph I, Moyer R A and Runov A M 2008 *Nucl. Fusion* **48** 024005
[13] Park G Y, Chang C S, Jeon Y M and Kim J Y 2012 Self-consistent kinetic simulation of RMP-driven transport: Collisionality and density effects on rmp penetration and transport *24th IAEA Fusion Energy Conference, October 8-13, 2012* (San Diego, USA) pp TH/P4–28
[14] Saad Y and Schultz M H 1986 *SIAM J. Sci. Stat. Comp.* **7** 856–869
[15] Renac F 2011 *J. Comp. Phys.* **230** 5739–5752
[16] Runov A M, Kasilov S V and Helander P 2015 *J. Comput. Phys.* **300** 605 – 622
[17] Kasilov S V, Runov A M and Kernbichler W 2016 *Comput. Phys. Commun.* **207** 282–286
[18] Lee J F, Sun D K and Cendes Z J 1991 *IEEE Trans. Microw. Theory Tech.* **39** 1262–1271
[19] Hecht F 2012 *J. Numer. Math.* **20** 251–265
[20] Liu Y, Kirk A and Nardon E 2010 *Phys. Plasmas* **17** 122502
[21] Ferraro N 2012 *Physics of Plasmas* **19** 056105

Journal of Physics: Conference Series **775** (2016) 012002 doi:10.1088/1742-6596/775/1/012002

Role of infernal modes dynamics and plasma rotation on the onset of NTMs in ECH-ECCD TCV plasmas

D Brunetti, E Lazzaro, S Nowak

Istituto di Fisica del Plasma IFP-CNR, Via R. Cozzi 53, 20125 Milano, Italy

O Sauter, J P Graves

École Polytechnique Fédérale de Lausanne (EPFL), Swiss Plasma Center (SPC), CH-1015 Lausanne, Switzerland

E-mail: `brunetti@ifp.cnr.it`

Abstract. Experiments in the TCV tokamak show that when high power central electron cyclotron heating (ECH) and current drive (ECCD) are deposited in the central plasma region, tearing-like instabilities can develop. In the present work we address, both analytically and numerically, the problem of tearing mode (TM) generation analysing extensively the behaviour of L-mode and low plasma current TCV discharges when high power heating is deposited in the plasma core region. Three possible mechanisms are discussed. The first is the modification of the equilibrium current density profile due to the power deposition leading to variation of the sign of the classical tearing stability index Δ', where finite β effects could also be taken into account. The second mechanism still the modification of the tearing stability parameter Δ', but this time due to the presence of an equilibrium sheared toroidal plasma rotation. Finally, the third mechanism considered is the role of *infernal-type* instabilities in driving tearing-like modes when the safety factor becomes sufficiently close to a rational in the core region.

1. Introduction

One of major issues in present day tokamak theory, is to understand the mechanisms which cause the onset of dangerous tearing-like instabilities. In absence of the standard tearing drive (i.e. current density gradient), neoclassical tearing modes (NTMs) can be driven unstable if *"seeded"* by other sources of magnetic perturbations, such as e.g. sawteeth and ELMs. In absence of these triggers, the onset of NTMs is still unexplained and it must be carefully investigated on the basis of the experimental evidence. Experimental observations in L-mode and low plasma current TCV discharges, showed the presence suddenly growing NTMs when high power heating is deposited in the plasma core region. This occurred without usual triggers such as sawteeth or ELMs [2] although a small 1/1 MHD activity was detected.

Hence this work we address the problem of NTM generation analysing a realistic TCV configuration when central power deposition and electron current drive (ECCD) are applied. We can think of three possible mechanisms for the triggering of these sudden tearing modes:

- a positive and possibly very large Δ' ($\Delta' = \psi'/\psi|_{r_s-\epsilon}^{r_s+\epsilon}$ [1]) due to the equilibrium current density profile modification
- a Δ' modification due to the effect of the equilibrium toroidal rotation

- a coupling between harmonics induced by the loss of field line bending in the core due to the flatness of the safety factor (infernal mechanism)

As proposed in Ref. [2], one of the driving mechanisms for the onset of these instabilities can be the modification of the equilibrium current density profile with a variation of the sign of the classical tearing stability index Δ'. Modifications of Δ' can be due also to finite β effect [3]. Indeed ECH and ECCD alter both the profile of the pressure, which is increased, and of the current viz. the safety factor which could be modified both at the position of q-rational surface of the observed NTM. These two effects combined could produce a change in Δ' leading to an effective classically unstable tearing mode. Modification of Δ' are also expected when the intrinsic equilibrium plasma rotation is taken into account [4].

From the equilibrium reconstruction and from the fact that there is a very small 1/1 MHD activity the safety factor appears to be shear-free over a fairly broad region in the core (up to $\rho_\psi \sim 0.35$) with $q \geq 1$. Configurations in which magnetic shear is allowed to become small over a large region, *infernal modes* can be driven unstable by the core pressure gradients if the core safety factor is sufficiently close to a rational number [5]. Infernal modes are characterised by toroidicity induced coupling between a core mode (with wave number (m, n)) and its neighbouring resistive sidebands (with wave number $(m \pm 1, n)$) which can grow on very fast timescales [6, 7] without the presence of usual triggers (e.g. sawteeth, ELMs). In particular we might expect co-current ECCD deposition should lower the central safety factor so that conditions for *infernal instability* are met. Hence infernal modes can be considered to be possible candidate to explain the onset of TMs observed experimentally.

Summarising, various mechanisms for the NTM generation are analysed having as a reference point a realistic TCV configuration with central ECH and ECCD. In section **2** we discuss the impact of the equilibrium sheared toroidal plasma rotation on Δ' assessing quantitatively its total modification with parameters typical for machines such as TCV, NSTX and DIII-D. Section **3** focuses on the numerical estimate of the classical Δ' (no rotation, classical TM). First we describe a novel code for the calculation of Δ', capable of handling realistic complex geometries without approximations on the metric elements. Then a series of four TCV discharges is analysed where the numerical computation of the tearing stability index is performed. Finally section **4** presents a parametric study of the resistive infernal stability in such configurations providing an estimate of the ideal threshold for these modes. Finally we summarise in section **5**.

2. Sheared rotation effects on Δ'

It is known that plasma rotation can have a significant effect on the nonlinear behaviour of a tearing mode [8]. But how big is the impact of the flow on the linear stability of a TM? The aim of this section is thus to investigate analytically the effects of toroidal plasma rotation on Δ'.

We set ourselves in a cylindrical coordinate system (r, ϑ, φ) where the magnetic field is given by

$$\boldsymbol{B} = \boldsymbol{\nabla} F \times \boldsymbol{\nabla}\vartheta - \boldsymbol{\nabla}\psi \times \boldsymbol{\nabla}\varphi. \tag{1}$$

The contravariant and covariant components of a vector $\boldsymbol{A}$ are given by $A^i = \boldsymbol{A} \cdot \boldsymbol{\nabla} q^i$ and $A_i = \boldsymbol{A} \cdot \boldsymbol{e}_{q^i}$, where $\boldsymbol{\nabla} q^i$ and $\boldsymbol{e}_{q^i}$ are the contravariant and covariant basis vectors respectively. We assume constant (both in space and time) mass density and the toroidal magnetic field. The equilibrium velocity is imposed to be $\boldsymbol{v}_0 = \Omega(r)\hat{\boldsymbol{e}}_\varphi$. In order to simplify the analysis we choose a reference frame which is moving along the axis of the cylinder such that $\Omega(r_s) = 0$ [9]. Let Ω_0 be the rotation frequency on the magnetic axis.

Multiplying the force balance equation by $1/B^\varphi$ and then taking the $\boldsymbol{\nabla}\varphi$ projection of its curl, leads to (for sake of simplicity we set $\Omega \equiv \Omega(r)$):

$$\frac{\varrho}{B^\varphi}\boldsymbol{\nabla}\varphi \cdot \boldsymbol{\nabla} \times (\boldsymbol{v} \cdot \boldsymbol{\nabla}\boldsymbol{v}) = \boldsymbol{B} \cdot \boldsymbol{\nabla}\left(\frac{J^\varphi}{B^\varphi}\right), \tag{2}$$

where finite β effects have been neglected.

A very simple Newcomb-like differential equation for the radial fluid displacement X is obtained by linearising (2) and the Faraday-Ohm's law with a parabolically distributed current [10]:

$$\frac{d}{dz}\Big[z^2[(1-z)^2-\hat{\Omega}^2]\frac{dX}{dz}\Big]+\Big\{\frac{(1-m^2)}{\lambda^2}[(1-z)^2-\hat{\Omega}^2]-2z\frac{\hat{\Omega}}{\lambda^2}\frac{d\hat{\Omega}}{dz}\Big\}X=0,$$

where the normalised variable $z=(r/r_s)^2$ has been introduced. The detailed analytic derivation is beyond the scope of this paper and it will be reported in a future publication. We limit ourselves to present the main results. For a parabolically distributed current density and toroidal rotation, the tearing stability parameter Δ' is found to be (see e.g. Ref. [10]):

$$r_s\Delta'=\frac{\bar{m}^2-m^2}{2}\pi\cot\pi\xi, \tag{3}$$

where $\xi=\frac{m-\bar{m}}{2}$ and $\bar{m}=\sqrt{m^2+8+4U}$. We defined the parameter $U=\frac{2[\Omega_0/(\hat{s}\omega_A)]^2}{1-[2\Omega_0/(\hat{s}\omega_A)]^2}$, where $\omega_A=B_0/(\sqrt{\varrho_0}R)$ is the Alfvén frequency and $\hat{s}=r_sq_s'/q_s$.

Assuming a magnetic shear of order of unity and $\Omega_0/\omega_A\sim10^{-2}\div10^{-3}$ [2, 8, 11], then Δ' is approximated by its usual cylindrical value plus a small correction which is found to be positive in agreement with Ref. [4], viz.:

$$r_s\Delta'\approx r_s\Delta'_{cyl}+\delta,$$

with $0<\delta\ll1$. We therefore conclude that rotation effects on the modification of the equilibrium Δ' are negligible.

3. Δ' estimate in ECH-ECCD TCV discharges

Before analysing the experimental TCV discharges, we describe the numerical code employed for the calculation of the stability index Δ'. This code works in a standard MHD frame where no equilibrium flows are considered, allowing in principle for finite β effects. Finite β corrections however are not taken into account in the analysis performed. We point out that in a realistic tokamak configuration shaping effects are not negligible. This is seen e.g. on the value of the edge safety factor in which the theoretical cylindrical model can differ up to the 30% wrt the real experimental value. Thus we developed our code in order to handle a generic tokamak geometry without any approximation.

The perturbed poloidal flux is calculated in a straight field line coordinate system (r,ϑ,φ) where the magnetic field is given by (1) and the radial variable r is normalised such that $r=1$ at the plasma boundary. The radial variable can be defined in terms of either the poloidal flux ($r\sim\sqrt{\frac{\psi-\psi_0}{\psi_{edge}-\psi_0}}$) or the toroidal flux ($r\sim\sqrt{\frac{F-F_0}{F_{edge}-F_0}}$) where ψ and F are as in (1) and the subscript 0 indicates that the quantity is evaluated on the magnetic axis. There is not a significant difference in the final results between the two definitions. Hence we decided to work with r defined in terms of the poloidal flux. In this coordinate system $\sqrt{g}$ denotes the Jacobian and $g_{i,j}$ $(i,j=r,\vartheta,\varphi)$ are the elements of the metric tensor.

Multiplying the force balance equation by $1/B^\varphi$ and then taking the $\nabla\varphi$ projection of its curl, after linearisation we have for the mth Fourier component of the perturbed flux (0 indicates equilibrium quantities):

$$\psi_m''=A_1\psi_m'+\Big(A_2+\frac{A_3}{m\mu-n}\Big)\psi_m, \tag{4}$$

where the Faraday-Ohm's law has been used, $\mu=1/q$, $J_*=J_0^\varphi/B_0^\varphi$, and the coefficients A_i $(i=1,2,3,4)$ are given by:

$$A_1=-\langle g_{\vartheta\vartheta}/\sqrt{g}\rangle^{-1}\langle g_{\vartheta\vartheta}/\sqrt{g}\rangle',\quad A_2=m^2\langle g_{\vartheta\vartheta}/\sqrt{g}\rangle^{-1}\langle g_{rr}/\sqrt{g}\rangle,\quad A_3=m\langle g_{\vartheta\vartheta}/\sqrt{g}\rangle^{-1}\langle J_*\rangle',$$

with $\langle A\rangle = \frac{1}{2\pi}\int_0^{2\pi} A d\vartheta$. In deriving the expression above we used the fact that $\langle g_{r\vartheta}/\sqrt{g}\rangle = \langle g_{r\vartheta}/\sqrt{g}\rangle' \sim 0$ and we neglected perturbations of the toroidal field. Note that no assumptions were made on the geometry of the system apart from the imposition of the vanishing average of the off diagonal elements of the metric tensor. We impose boundary conditions $\psi_m(0) = \psi_m(1) = 0$.

The code reads an OUTXOR-CHEASE equilibrium [12], viz. pressure and current profiles and the shaping of the magnetic surfaces. The CHEASE equilibrium needs as input plasma pressure and the toroidal current I^* giving as output the safety factor, required for the stability calculation. A recent update of the code [13] allows to provide as input the safety factor. This new feature is not used in the preset work but it will be useful in further studies for a simpler and more rapid estimate of Δ'.

First the straight field line poloidal angle is calculated (see e.g. [14]). The the metric tensor elements $g_{i,j}$ and their averages are calculated by Fourier decomposing the magnetic surfaces in ϑ. In this manner the calculation of the radial and poloidal derivatives (of e.g. the toroidal current) becomes straightforward. Once we have all the physical quantities, the coefficients A_i $(i = 1,4)$ can be evaluated.

We note that for $m = 2$ the evaluation of Eq. (4) becomes problematic since the coefficient A_1 is singular on the magnetic axis while $\psi''_m \neq 0$ (this is clear because near the magnetic axis the perturbed poloidal flux is expected to behave as $\psi_m \sim r^m$). Hence when $m = 2$, equation (4) is solved in the region $0 < r < r_s$ with a RK4 method with the following initial conditions at r_0 near the magnetic axis (Cauchy problem):

$$\psi_m \sim r_0^m, \quad \psi'_m \sim m r_0^{m-1} \tag{5}$$

When $m \geq 3$ we have that $\psi''_m = 0$ on the magnetic axis so that the shooting method implemented with an RK4 procedure can be employed in which we "shoot" the solution either from the magnetic axis to the resonant point or the other way around. The shooting method (away from r_s to the plasma boundary) is used in the region $r_s < r < 1$ for any m since in this region the coefficients A_i are well defined. We note that instead of using a shooting procedure for solving ψ_m when $m \geq 3$ in the region $0 < r < r_s$, we can define an analogous Cauchy problem as already done for the $m = 2$ case (cf. Eq. 5). The final result is practically identical to the one obtained with the shooting method, hence in our calculation we always (i.e. for any m) assume a Cauchy problem of the type (5) in the region $0 < r < r_s$. In order to increase the accuracy of the evaluation of the tearing eigenfunction the radial mesh becomes finer as the resonant surface of the mode is approached.

Finally, ψ'_m is fitted with a logarithmic least square fitting algorithm with a function of the type $\psi'_m \sim a_L + b_L \ln(x)$ for $x < 0$ and $\psi'_m \sim a_R + b_R \ln(x)$ for $x > 0$, where $x = r - r_s$ and $\Delta' = a_R - a_L$.

Excellent agreement is found in the large aspect ratio limit with the theoretical results given in Ref. [3]. As a preliminary result, we have also implemented finite β effects. When β effects are switched off Δ' reduces to the zero pressure case. In addition, in line with Ref. [15], we checked that $\Delta' \to -2$ as $D_I \to \frac{1}{4}$ (i.e. pressure increased), which guarantees the well behaviour of the algorithm. However since a more refined modelling of the finite pressure effects is required, β corrections will not be considered when we discuss the application of the code for the analysis of realistic TCV discharges. This will be presented in the next section.

3.1. Analysis of TCV discharges #488(36),(37),(39),(41)

We analyse four TCV shots, namely #488(36),(37),(39),(41). The two key parameters of interest for our study are the pressure profile and the shape of the safety factor, i.e. the plasma current density. These two quantities are modified during time both due to central ECH and ECCD.

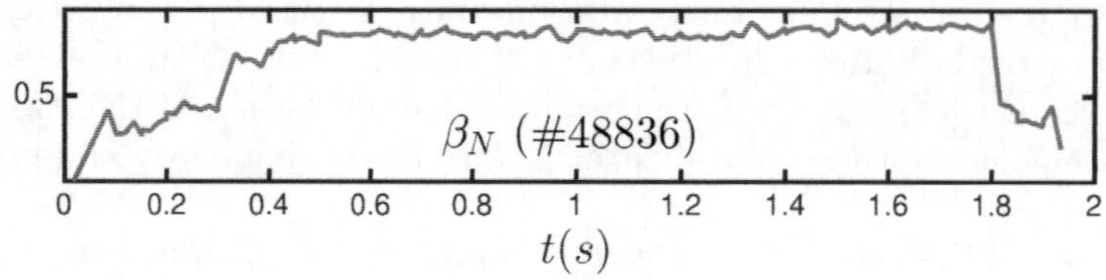

Figure 1. Time evolution of β_N for shot#48836. A similar behaviour is also observed for shots #488(37),(39),(41). The time trace of the profile remains qualitatively and quantitatively similar, despite MHD activity is present (cf. Fig. 2).

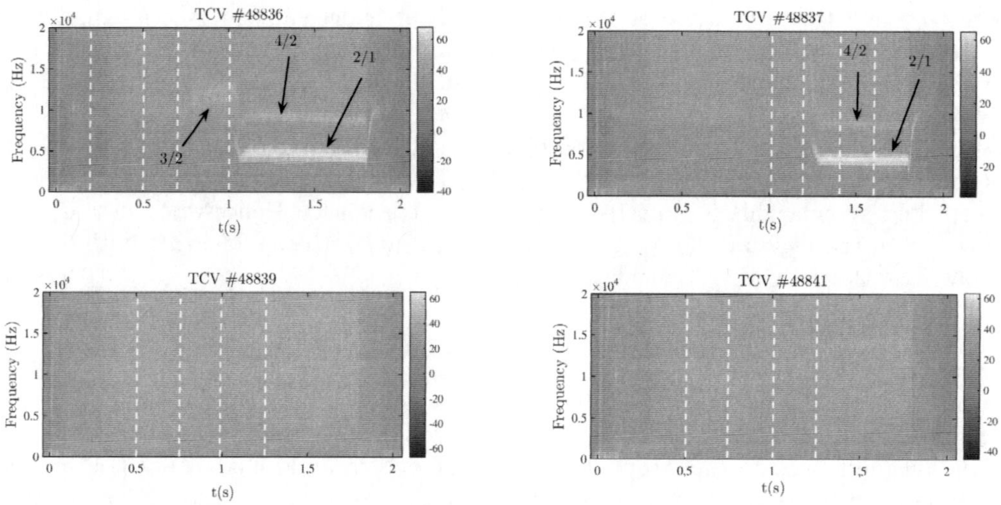

Figure 2. Spectrograms of TCV discharges #488(36),(37),(39),(41). No MHD activity is recorded in shots #488(39),(41), where the current injection was counter-current. Shots #488(36),(37) with co-current injection show a low m MHD activity: modes 3/2 and 2/1 are observed in shot #48836 while only the 2/1 is observed in shot #48837. The dashed white vertical lines correspond to the times at which the safety factor profiles shown in figure 3 are taken.

The current injection is mainly in the co-I_P (I_P is the plasma current) direction for discharges #488(36),(37) and mainly in counter-I_P direction for shots #488(39),(41). The ECH phase starts in the four discharges at $t \sim 0.2 - 0.3s$. This causes a sudden rise of the pressure as it is clearly visible in the time trace of the evolution of $\beta_N(= \beta a B_T / I_P$, where a is the minor radius and B_T is the toroidal magnetic field on the axis) shown in Fig. 1. No significant variations in β_N are observed as the 3/2 and 2/1 TMs appear. Figure 2 shows the spectrograms for the four discharges analysed. We see that MHD activity is present in shots #488(36),(37) (co-I_P ECCD) while discharges in #488(39),(41) no TM activity is reported.

In order to evaluate Δ', we need a CHEASE equilibrium which is obtained from the instantaneous equilibrium reconstruction via the LIUQE code [16]. Once the CHEASE equilibrium is calculated we are able to compute numerically the instantaneous tearing stability index Δ' with the code described in the previous section.

The safety factor profiles taken at the times indicated in Fig. 2 are shown in figure 3. First

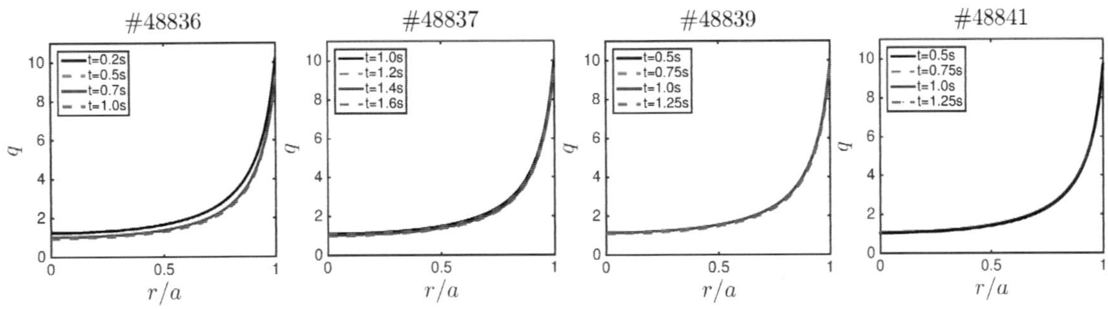

Figure 3. Radial safety factor profiles taken at different times for the shots of the series #488(XX) considered. Note that there are no dramatic variations at the 2/1 and 3/2 resonant surfaces.

of all we note that the MHD activity does not appear immediately after the ECH and ECCD deposition. This might suggest that the mechanism of the onset of the TMs observed should be contained in the variations of the current density profile. From figure 3, we note that in the four shots the variations q during time after the ECH deposition phase are not dramatic (e.g. at the 2/1 resonant surface). We however may expect substantial differencences in q'', leading to different values of the stability parameter Δ'. Hence we can numerically calculate Δ' for the Fourier modes 2/1, 3/2, 3/1, 4/2 and 4/3. This is shown in figure 4 where finite β effects have been dropped. It is observed that when $\beta = 0$, the 2/1 mode has always a positive Δ', contrarily to what we expect from the experimental observation for shots #488(39),(41) where this mode is stable which would imply $\Delta' < 0$. In addition we note that in shots #488(36),(37) Δ' of the mode 2/1 is found to be positive well before the onset of the instability. This could be an indication that a more subtle underlying physics has to be considered. It is however found that modes with $m \geq 3$ are stable ($\Delta' < 0$) in all four shots so that the mechanism of a simple modification of the current profile (i.e. q) at the mode resonant surface is not sufficient for explaining the onset of the 3/2 mode observed in discharge #48836. In the next section we discuss the third mechanism which could explain this behaviour.

4. Role of the infernal mode mechanism for the TM onset

As shown in figure 1, in all four shots β_N does not have significant variations when the resistive modes appear. In particular the four discharges analysed have almost the same pressure profile. Thus we might expect to be unlikely that the MHD activity reported depends solely on the rise of the pressure due to the ECH. As shown in figure 5, the CHEASE reconstructed q profile is very close to $q = 1$ over a rather large region which extends from the magnetic axis to $r/a \sim 0.3$. We recall that shots #488(36),(37) were performed globally in co-I_p ECCD (3 gyrotrons in co-I_p for #48836 and 2 gyrotrons in co-I_p and 1 in counter-I_p for #48837) with a consequent increase of the total I_P. Conversely shots #488(39),(41) were globally in counter-I_p ECCD (2 gyrotrons in counter-I_p and 1 in co-I_p in #48839 and 1 gyrotron in counter-I_p, 1 in co-I_p and 1 gytron in heating in #48841) with a decrease of I_p. It is known that an increase of the total current leads to a small decrease of q, i.e. the safety factor is shifted downwards almost rigidly [17]. On the other hand a decrease of the total current should "*lift*" the q.

This might suggest that an infernal-like mechanism could occur in which a core 1/1 (2/2) mode couples with its 2/1 (3/2) tearing sideband. In particular small 1/1 activity was measured in shot #48836 although invisible on the spectrograms in figure 2 (this could be due to the fact that these are internal modes with a not too large amplitude). Infernal modes are more

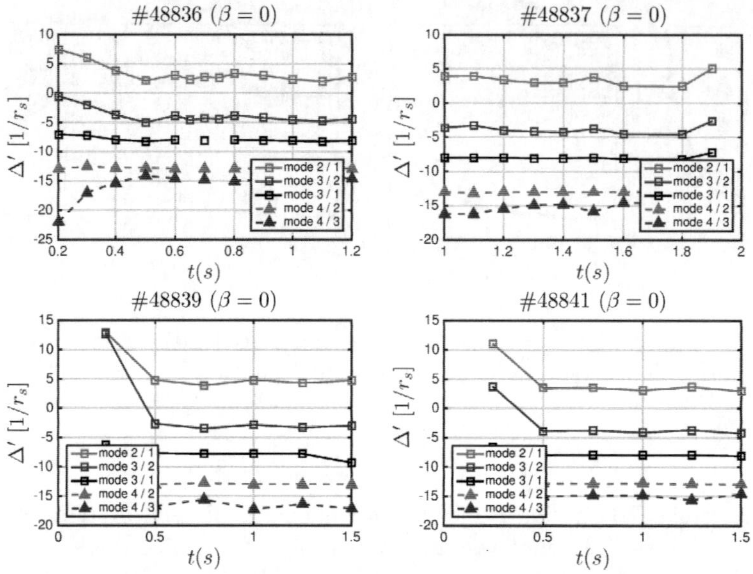

Figure 4. Values of Δ' computed numerically for discharges #488(36),(37),(39),(41) where the modes 2/1,3/2,3/1,4/2 and 4/3 are considered under the assumption $\beta = 0$. Although no MHD signals were observed in shots #488(39),(41) as shown in Fig. 2, the code still gives a positive definite Δ' for the 2/1 mode which appears to contradict the experimental observation.

unstable if the core safety factor is close to the rational and are stabilised as q is pushed away. This could explain why in discharges #488(36),(37) MHD activity is observed while being absent in shots #488(39),(41). This mechanism thus appears to be plausible and not in contrast with the experimental observations. We focus our analysis on the 3/2 mode since the 2/1 mode was numerically found to be unstable requiring further study to find an agreement with experimental observations.

We know that if the tearing stability index is negative, a threshold for the onset of the resistive infernal mode in the value of q_{min} for a fixed β_p is expected [5]. This means that by writing $q = 1 + \delta q$ it does exist a critical value δq_{crit} so that if $\delta q < \delta q_{crit}$ the coupling occurs and the resistive sideband is unstable. If $\delta q > \delta q_{crit}$ the resistive sideband is stabilised. The same argument hold when we swap δq with β_P, i.e. we fix δq and let β_P to vary. Of course in this case the threshold is given by a critical value of poloidal β (i.e. $\beta_{P,crit}$) so that if $\beta_P > (<)\beta_{P,crit}$ the mode is unstable(stable). This in principle could explain the sudden appearance of the tearing instability as observed in Fig. 2.

Expressing the safety factor in the central region where it is almost flat as $q \approx 1 + \delta q$, the threshold is found by setting $\gamma = 0$ (γ is the growth rate) and gives the following equation for a fixed poloidal mode m:

$$\delta q = \sqrt{-\beta_{p,*}^2 G_0 \frac{M_0}{A_0}}, \tag{6}$$

where β_p is the poloidal β and it is evaluated at the transition radius r_* where the magnetic shear $\hat{s} = (rdq/dr)/q$ becomes smaller than unity, i.e. where $\hat{s}(r = r_*) \approx 1$ (all the quantities with the subscript $*$ are evaluated at the point $r = r_*$). We defined $\varepsilon = r/R_0$, $G_0 = \frac{\varepsilon_*^2 \hat{s}_*/(\hat{s}_* - 2)}{m(m+1)}$, A_0 and M_0 are a measure of the amplitude of the jump of the eigenfunction across the resonant surface

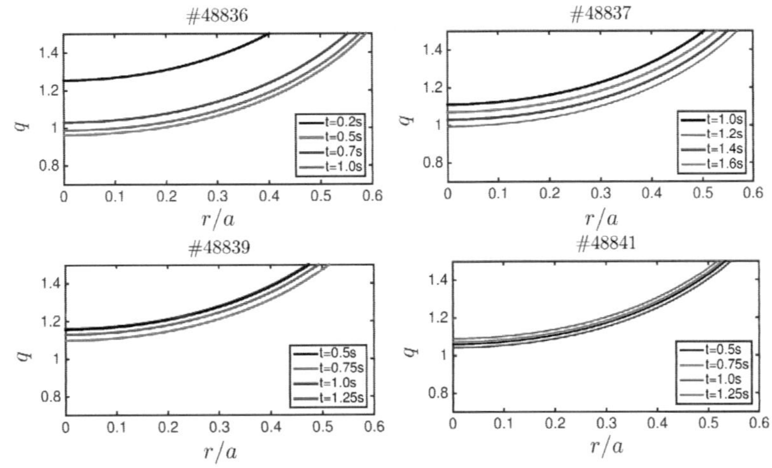

Figure 5. Radial safety factor profiles obtained from CHEASE equilibrium reconstruction taken at different times for the shots of the series #488(XX) considered (times taken as in Fig. 3). We note that for shots #488(36),(37) the central q is closer to unity compared with discharges #488(39),(41). This might suggest that an infernal-like mechanism is plausible for the triggering of the modes.

of the mode (A_0 is the standard tearing Δ' and M_0 is the jump of the second eigenfunction which is usually dropped because of the divergence on the magnetic axis).

Letting A_0 and M_0 vary (usually in the cylindrical limit $M_0 < 0$) and assuming a stable mode with $A_0 < 0$ we obtain the threshold value for δq_{crit} below which the mode becomes unstable. This is reported in Fig. 6, where we used the experimental values of TCV shot #48836 for β_P, ε and $\hat{s}$ for the 3/2 mode for which $\Delta' < 0$ (cf. Fig. 4). In this case we have $r_*/a \approx 0.3$. We find $\delta q_{crit} \sim 0.1$ which is consistent with the shape of the safety factor shown in Fig. 5. Thus an infernal-like mechanism could be invoked as a valid candidate for the explanation of the onset of the 3/2 mode in TCV discharge #48836.

5. Conclusions

Three possible mechanisms for the explanation of the onset of TMs in realistic TCV discharges have been analysed. First the role of the equilibrium intrinsic toroidal rotation has been assessed. It is found that for realistic values of the toroidal rotation (valid for TCV, DII-D and NSTX tokamaks), the value of Δ' is only weakly increased leading us to rule out such effect as a possible trigger for the sudden appearance of TMs.

The second mechanism analysed is the modification of the safety factor profile, i.e. current density, due to the combined action of central ECH and ECCD deposition. A modification of the current density profiles leads to a variation of Δ'. In order to evaluate numerically Δ', a novel code has been developed which allows an accurate determination of the tearing stability index without any approximation on the geometry. This code has been used to analyse four TCV discharges (#488(36),(37),(39),(41)). It has been found that modes with $m \geq 3$ are stable ($\Delta' < 0$) while the $m = 2$ always presents a positive Δ'. This cannot explain such a sharp rise of TMs and suggests that a more subtle physics has to be unravelled.

Finally the role of the infernal mechanism for the 3/2 mode has been assessed. It is found that in the range of parameters typical of the four TCV shots analysed, the threshold of the resistive mode is not too far from the experimental evidence. Moreover this mechanism appears to be

Journal of Physics: Conference Series **775** (2016) 012002 doi:10.1088/1742-6596/775/1/012002

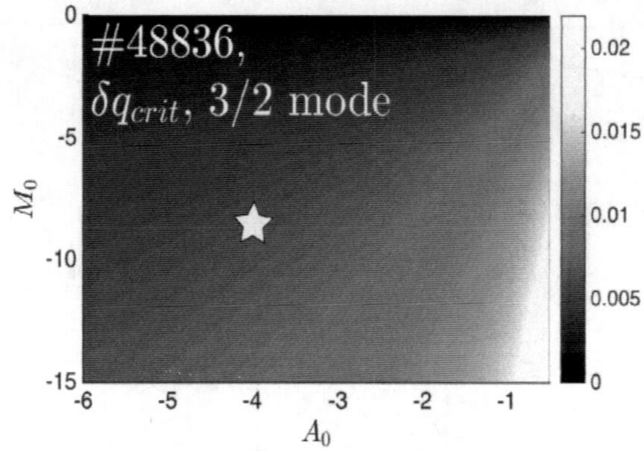

Figure 6. Threshold in δq for the mode 3/2 with the safety factor and β_p of the shot #48836. As the mode becomes more stable (decreasing A_0) δq_{crit} decreases. This means that instability window $\delta q < \delta q_{crit}$ for which the mode is unstable becomes narrower, i.e. q has to be very close to unity to drive and instability and if it lies outside this window no mode is expected. Experimental values of A_0, M_0 for shot #48836 are indicated by the star.

plausible and not in contrast with the experimental observations since resistive MHD activity was reported only in shots with co-I_P ECCD, i.e. when infernal unstable conditions are met.

Further future investigation will focus on a more refined description of the finite β effects and in particular analysing the nonlinear behaviour described by the generalised Rutherford equation taking into account possible variations in the sign of the polarisation term in the generalised Rutherford equation [2, 18]). It is envisaged that this work will guide the nonlinear work to follow.

Acknowledgments

This work, performed under the European Fusion Development Agreement, was supported by the European Communities, the Istituto di Fisica del Plasma CNR and the Swiss National Science Foundation. The views and opinions expressed herein do not necessarily re ect those of the European Commission.

References

[1] Wesson J A 2011, **Tokamaks** (Oxford U. P., Oxford, UK)
[2] Lazzaro E et al. 2015 *Nucl. Fusion* **55** 093031
[3] Hegna C C and Callen J D 1994 *Phys. Plasmas* **1** 2308
[4] Sen A, Chandra D and Kaw P 2013 *Nucl. Fusion* **53** 053006
[5] Brunetti D, Graves J P, Cooper W A and Wahlberg C 2014, *Plasma Phys. Control. Fusion* **56** 075025
[6] Charlton L A, Hastie R J and Hender TC 1989 *Phys. Fluids* B **1** 798
[7] Brunetti D, Graves J P, Halpern F D, Luciani J-F, Lütjens H and Cooper W A 2015, *Plasma Phys. Control. Fusion* **57** 054002
[8] La Haye R J, Brennan D P, Buttery R J and Gerhardt S P 2010 *Phys. Plasmas* **17** 056110
[9] Chen X L and Morrison J P 1990 *Phys. Fluids B* **2** 495
[10] Mikhailovskii A B 1998 *Instabilities in a Confined Plasma* (IOP, Bristol)
[11] Wang F, Fu G Y, Breslau J A, Tritz K and Liu J Y 2013 *Phys. Plasmas* **20** 072506
[12] Lütjens H, Bondeson A and Sauter O 1996 *Comput. Phys. Commun* **97** 219
[13] Blondel L 2016 *Axisymmetric Tokamak Equilibria computed with a Predefined Safety Factor Profile as CHEASE input* (EPFL Master Project)
[14] Hirshman S P and Whitson J C 1983 *Phys. Fluids* **26** 3553

[15] Nishimura Y, Callen J D and Henga C C 1998 *Phys. Plasmas* **5** 4292
[16] Moret J-M, Duval B P, Le H B, Coda S, Felici F and Reimerdes H 2015 *Fusion Eng. Des.* **91** 1
[17] Cooper W A, Graves J P and O Sauter O 2011 *Plasma Phys. Control. Fusion* **53** 024002
[18] Smolyakov A I, Hirose A, Lazzaro E, Re G B and Callen J D 1995 *Phys. Plasmas* **2** 1581

Journal of Physics: Conference Series **775** (2016) 012003 doi:10.1088/1742-6596/775/1/012003

Non-Maxwellian background effects in gyrokinetic simulations with GENE

A. Di Siena[1], T. Görler[1], H. Doerk[1], J. Citrin[2,3], T. Johnson[4], M. Schneider[3], E. Poli[1] and JET Contributors[5]

[1]Max Planck Institute for Plasma Physics, Boltzmannstr.2, 85748 Garching, Germany
[2]FOM Institute DIFFER, PO Box 6336, 5600 HH Eindhoven, The Netherlands
[3]CEA, IRFM, F-13108 Saint Paul Lez Durance, France
[4]VR Association, EES, KTH, Stockholm, Sweden
[5]EUROfusion Consortium, JET, Culham Science Centre, Abingdon, OX14 3DB, UK

E-mail: adisi@ipp.mpg.de

Abstract. The interaction between fast particles and core turbulence has been established as a central issue for a tokamak reactor. Recent results predict significant enhancement of electromagnetic stabilisation of ITG turbulence in the presence of fast ions. However, most of these simulations were performed with the assumption of equivalent Maxwellian distributed particles, whereas to rigorously model fast ions, a non-Maxwellian background distribution function is needed. To this aim, the underlying equations in the gyrokinetic code GENE have been re-derived and implemented for a completely general background distribution function. After verification studies, a previous investigation on a particular JET plasma has been revised with linear simulations. The plasma is composed by Deuterium, electron, Carbon impurities, NBI fast Deuterium and ICRH ^{3}He. Fast particle distributions have been modelled with a number of different analytic choices in order to study the impact of non-Maxwellian distributions on the plasma turbulence: slowing down and anisotropic Maxwellian. Linear growth rates are studied as a function of the wave number and compared with those obtained using an equivalent Maxwellian. Generally, the choice of the ^{3}He distribution seems to have a stronger impact on the microinstabilities than that of the fast Deuterium.

1. Introduction

Developing a better understanding of the impact of fast particles on plasma core turbulence is an essential task for an improved performance assessment of future fusion reactors like ITER, where the energetic particles fraction is significant. In this work, the interaction between fast particles and plasma core turbulence is studied in the radially local limit. Therefore, energetic geodesic acoustic mode (EGAMs) effects will not be taken into account. However, the latter are usually negligible in the core plasma region [1]. In this context, recent experimental and computational results suggest positive and stabilising effects on the background turbulence, especially on the ITG modes [2, 3], typically considered as the most limiting microinstability in a tokamak reactor. In particular, fast ion dilution of the main ion species [4, 5], Shafranov shift stabilisation [6] as well as linear and nonlinear electromagnetic stabilisation in the presence of fast ions suprathermal pressure gradients [2, 7] have been proved to greatly stabilise the plasma core turbulence. Although some previous gyrokinetic studies could already well reproduce the experimental observation, the linear reduction of the growth rates and the nonlinear reduction of the main ion heat flux are somewhat too strong such that the power balance was reached with R/L_{T_i} above the experimental values [8]. However, most of these simulations have been performed with the assumption of equivalent Maxwellian distributed particles, whereas it is well known that to rigorously model fast particles, a non-Maxwellian background distribution function is needed. Indeed, anisotropies and asymmetries in the fast particles distribution functions can, in principle, have some additional effects on the main plasma turbulence. For this reason, in this paper we present low beta electromagnetic simulations with non-Maxwellian fast particles distribution functions with the δf gyrokinetic code GENE [9]. The latter has been modified and generalised in order to include any number of analytic non-Maxellian fast particles species, namely slowing down and asymmetric and anisotropic Maxwellian, which as already mentioned, provide the most appropriate description. Linear electrostatic benchmarks against published results [10] obtained with GKW [11, 12] and GS2 [13, 14] are presented, which include slowing down fast particles in the electrostatic case. Furthermore, a relatively low beta JET plasma with both NBI and ICRH fast ions has been revised and analysed with linear simulations. Comparisons on the linear growth rates are shown in this paper between the different fast ion distribution functions and the previous Maxwellian results.

2. Theoretical background

2.1. Equilibrium distribution functions

All the simulations presented in this work have been carried out with the gyrokinetic δf code GENE, which solves numerically the nonlinear coupled Vlasov-Maxwell system of equations on a five dimensional grid for each time step. GENE can include full electromagnetic effects, realistic collisional operators as well as experimental geometries. The background distribution function usually considered in most of the gyrokinetic codes is a local Maxwellian (defined in Eq.(1)). The latter is a very good approximation for all the thermalised species, however, it lacks validity for fast particles where anisotropies and asymmetries cannot be described with a Maxwellian distribution function. In order to describe appropriately the dynamic of non thermalised particles, the gyrokinetic equations have been re-derived and implemented without doing any assumption on the background distribution function in the low-beta electromagnetic case, i.e. neglecting the parallel magnetic fluctuations ($B_{1,\parallel}$). This very flexible setup allows to implement and use a large variety of different distribution functions and study the impact of energetic particles on the plasma turbulence. In the present work the equilibrium distribution function F_0 for thermal electrons and ions has been modelled with a Maxwellian

$$F_{0,M} = \frac{n_0}{\pi^{3/2} v_{th}^3} \exp\left(\frac{-m v_\parallel^2/2 - \mu B_0}{T_0}\right) \tag{1}$$

Here, m is the particle mass, T_0 is the equilibrium temperature, n_0 particles density and $v_{th} = (2T_0/m)^{1/2}$ is the thermal velocity. The fast particle distributions have been modelled with a number of different analytic choices: slowing down [15] and an asymmetric and anisotropic Maxwellian [16]. The slowing down is a solution of the Fokker-Planck equation with an isotropic delta-function particle source and is defined as follow

$$F_{0,s} = \frac{3n_0}{4\pi \log\left(1 + \frac{v_\alpha^3}{v_c^3}\right)\left[v_c^3 + v^3\right]} \Theta\left(v_\alpha - v\right) \tag{2}$$

Here, the birth velocity is defined through the birth energy E_α in the following way $v_\alpha = (2E_\alpha/m_\alpha)^{1/2}$, while $v_c = v_{th,e}\left(\frac{3\sqrt{\pi} m_e}{4} \sum_{\text{main ions}} \frac{n_i z_i^2}{n_e m_i}\right)^{1/3}$ represents the critical slowing down velocity. Furthermore, Θ is the Heaviside step function. The asymmetric and anisotropic

Maxwellian distribution function is defined as follows

$$F_{0,aM} = \frac{n_0 \exp\left(-\frac{\mu B_0}{T_0}\right)}{\pi^{3/2}\left(v_{th,-}+v_{th,+}\right) v_{th,\perp}^2}\left\{\exp\left(-\frac{v_{\shortparallel}^2}{v_{th,-}^2}\right)\left[1-\Theta\left(v_{\shortparallel}\right)\right]+\exp\left(-\frac{v_{\shortparallel}^2}{v_{th,+}^2}\right)\Theta\left(v_{\shortparallel}\right)\right\} \quad (3)$$

Here, $T_{\shortparallel,+}$ and $T_{\shortparallel,-}$ are respectively the temperature associated to the particles with positive and negative parallel velocity, while $T_\perp$ is the perpendicular temperature. In the limit of $T_{\shortparallel,+} = T_{\shortparallel,-}$ Eq.(3) reduces to the bi-Maxwellian distribution function, which is often used to model ICRH fast ions. The latter will be used afterwards for modelling the fast ICRH ions in the linear simulations. Furthermore, if $T_{\shortparallel,+} = T_{\shortparallel,-} = T_\perp$ Eq.(3) turns into Eq.(1). The non-Maxwellian fast particles temperature has been defined in order to have the same kinetic pressure as the equivalent Maxwellian distribution function [17], namely

$$\int v^2 F_{non-M} dv = \int v^2 F_M dv$$

In the following session the derivation of the non-Maxwellian gyrokinetic equations is presented in the low-beta electromagnetic case.

2.2. Non-Maxwellian gyrokinetic Vlasov-Maxwell equations

In the present section the analytic derivation of the non-Maxwellian gyrokinetic equations is revised in the low-beta electromagnetic case without doing any assumptions on the background distribution function of the plasma species. As starting point, we write the Vlasov equation which determines the time evolution of the gyrocenter distribution function [18] $F = F\left(\vec{X}, v_{\shortparallel}, \mu\right)$

$$\frac{\partial F}{\partial t} + \left[v_{\shortparallel}\hat{b}_0 + \left(\vec{v}_E + \vec{v}_{\nabla B} + \vec{v}_c\right)\right]\cdot\left\{\vec{\nabla}F - \left[q\vec{\nabla}\bar{\phi}_1 + \frac{q}{c}\hat{b}_0\dot{\bar{A}}_{1,\shortparallel} + \mu\vec{\nabla}B_0\right]\frac{1}{mv_{\shortparallel}}\frac{\partial F}{\partial v_{\shortparallel}}\right\} = 0 \quad (4)$$

Here, respectively the curvature, the $E \times B$ and the ∇B drift velocities have been defined $\vec{v}_C = \frac{v_{\shortparallel}^2}{\Omega}\left(\vec{\nabla}\times\hat{b}_0\right)_\perp$, $\vec{v}_E = \frac{c}{B_0^2}\left(\vec{B}_0 \times \vec{\nabla}\xi_1\right)$ and $\vec{v}_{\nabla B_0} = \frac{\mu c}{q B_0^2}\left(\vec{B}_0 \times \vec{\nabla}B_0\right)$; ξ_1 represents the modified potential $\xi_1 = \bar{\phi}_1 - \frac{v_{\shortparallel}}{c}\bar{A}_{1,\shortparallel}$; $\Omega = \frac{qB_0}{m}$ and $\hat{b}_0 = \frac{\vec{B}_0}{B_0}$. The over bar denotes gyro averaged quantities, which in the local code approximation reduces to the mere multiplication of the zero order Bessel function, i.e. $\bar{\phi}_1 = J_0(\lambda)\phi_1$; $\bar{A}_{1,\shortparallel} = J_0(\lambda) A_{1,\shortparallel}$; where $\lambda = \frac{k_\perp}{\Omega}(\frac{2B_0\mu}{m})^{1/2}$. In terms of computational effort it is convenient to split the gyrocenter distribution function in an equilibrium part F_0 and a small perturbation term F_1, with $F_1 \ll F_0$. The turbulent evolution of the system is then determined within the so-called δf-ordering of Eq.(4) which for a field aligned coordinate system can be rewritten in terms of the metric coefficients $g^{ij} = \nabla u^i \cdot \nabla u^j$ (with $u^i = (x, y, z)$, x radial direction, y poloidal direction, z toroidal direction) in the following way

$$\frac{\partial g_1}{\partial t} + \frac{C}{JB_0}\left\{v_{\shortparallel}\partial_z F_1 - \left(\frac{q}{m}\partial_z\phi_1\frac{\partial F_0}{\partial v_{\shortparallel}} + \frac{\mu}{m}\partial_z B_0 \frac{\partial F_1}{\partial v_{\shortparallel}}\right)\right\} +$$

$$+\frac{cB_0}{CB_{o,\shortparallel}^*}\left(\frac{g^{1i}g^{2j} - g^{2i}g^{1j}}{\gamma_1}\right)\left\{\left[\partial_i\xi_1 + \frac{\mu}{q}\partial_i B_0 + \frac{v_{\shortparallel}^2 m}{q}\left(\frac{\partial_i B_0}{B_0} + \frac{\beta_p}{2}\frac{\partial_i p_0}{p_0}\right)\right]\right.$$

$$\left.\left[\partial_j F_0 + \partial_j F_1 - \left(q\partial_j\phi_1 + \mu\partial_j B_0\right)\frac{1}{mv_{\shortparallel}}\frac{\partial F_0}{\partial v_{\shortparallel}}\right]\right\} = 0 \quad (5)$$

a modified distribution function $g_1 = F_1 - \frac{q}{mc}\bar{A}_{1,\shortparallel}\frac{\partial F_0}{\partial v_{\shortparallel}}$ has been introduced. C is defined below. In order to solve numerically Eq.(5) we need to derive a dimensionless equation. With this aim,

Journal of Physics: Conference Series **775** (2016) 012003 doi:10.1088/1742-6596/775/1/012003

all the physical quantities have been split into a dimensionless value and a dimensional reference part. The reference value used to normalise Eq.(5) are the elementary electron charge e, the main ions mass m_i and temperature T_i, a reference magnetic field B_{ref} and a macroscopic length L_{ref}. The normalised Vlasov equation for a completely general background distribution function can be written as

$$\begin{aligned}
\frac{\partial g_1}{\partial t} &= -\frac{C}{JB_0} v_{th} v_{\shortparallel} \left[\partial_z F_1 - \frac{q}{2T_0 v_{\shortparallel}} \partial_z \phi_1 \frac{\partial F_0}{\partial v_{\shortparallel}} \right] + \frac{C}{JB_0} v_{th} \frac{\mu}{2} \partial_z B_0 \frac{\partial F_1}{\partial v_{\shortparallel}} + \\
&+ \frac{T_0}{q} \left(\frac{\mu B_0 + m v_{\shortparallel}^2}{B_0} \right) \mathcal{K}_x \hat{\partial}_x F_0 - \frac{T_0}{q} \left[\left(\frac{\mu B_0 + m v_{\shortparallel}^2}{B_0} \right) \mathcal{K}_y - \frac{1}{C} \frac{v_{\shortparallel}^2 \beta_{ref}}{B_0^2} \omega_p \right] \partial_y g_1 + \\
&+ \left[\frac{1}{2v_{\shortparallel}} \frac{\partial F_0}{\partial v_{\shortparallel}} \left(\frac{\mu B_0 + m v_{\shortparallel}^2}{B_0} \right) \mathcal{K}_y - \frac{1}{C} \frac{v_{\shortparallel}^2 \beta_{ref}}{B_0^2} \omega_p \frac{1}{2v_{\shortparallel}} \frac{\partial F_0}{\partial v_{\shortparallel}} - \frac{1}{C} \hat{\partial}_x F_0 \right] \partial_y \xi_1 + \\
&- \frac{T_0}{q} \left(\frac{\mu B_0 + m v_{\shortparallel}^2}{B_0} \right) \mathcal{K}_x \partial_x g_1 + \frac{1}{2v_{\shortparallel}} \frac{\partial F_0}{\partial v_{\shortparallel}} \left(\frac{\mu B_0 + m v_{\shortparallel}^2}{B_0} \right) \mathcal{K}_x \partial_x \xi_1 + \\
&- \frac{1}{C} \left[\partial_x \xi_1 \partial_y g_1 - \partial_y \xi_1 \partial_x g_1 \right]
\end{aligned} \tag{6}$$

where we have defined the following geometrical coefficients $\mathcal{K}_x = -\frac{1}{C}\left(\partial_y B_0 - \frac{\gamma_3}{\gamma_1}\partial_z B_0\right)$, $\mathcal{K}_y = \frac{1}{C}\left(\partial_x B_0 - \frac{\gamma_3}{\gamma_1}\partial_z B_0\right)$, $\gamma_1 = g^{11}g^{22} - g^{21}g^{12}$, $\gamma_3 = g^{12}g^{23} - g^{22}g^{13}$ and $C = B_0/\gamma_1^{1/2}$; the normalised background pressure gradient $\omega_p = -\mathcal{L}_{\text{ref}} \frac{\partial_x p_0}{n_{\text{ref}} T_{\text{ref}}}$ and the reference thermal to magnetic pressure ratio $\beta_{\text{ref}} = \frac{8\pi n_{\text{ref}} T_{\text{ref}}}{B_{\text{ref}}^2}$. If the equilibrium distribution function F_0 is chosen to be a local Maxwellian, one can show that Eq.(6) reduces to the well-known gyrokinetic equation known in literature see, e.g., Ref.[18].

2.3. Velocity space moments

In order to treat self-consistently our fluctuating system it is necessary to derive the fluctuating component of the electromagnetic fields from the perturbed distribution function. In this section the evaluation of the general moment of the perturbed distribution function

$$M_{a,b}(\vec{x}) = \int f(\vec{x}, \vec{v})\, v_{\shortparallel}^a v_{\perp}^b\, d^3v$$

is established. Since, from the Vlasov equation we can evaluate the time evolution of the gyrocenter distribution function it is convenient to transform the integrand of the general velocity moment equation in terms of the gyrocenter coordinates:

$$M_{a,b}(\vec{x}) = \int \delta\left(\vec{X} + \vec{r} - \vec{x}\right) T^* F_1 \frac{B_0\left(\vec{X}\right)}{m} v_{\shortparallel}^a v_{\perp}^b\, d^3X\, dv_{\shortparallel}\, d\mu\, d\theta$$

where $\vec{r}$ is the gyroradius vector and T^* is the pull back operator, which acts on the perturbed distribution function in the following way

$$\begin{aligned}
T^* F_1 = F_1 + \frac{1}{B_0} \Bigg\{ &\left(\Omega \frac{\partial F_0}{\partial v_{\shortparallel}} - \frac{q}{c} v_{\shortparallel} \frac{\partial F_0}{\partial \mu} \right) \cdot \left(\vec{A}_1\left(\vec{X} + \vec{r}\right) - \bar{\vec{A}}_1\left(\vec{X}\right) \right) \\
&+ q\left(\phi_1\left(\vec{X} + \vec{r}\right) - \bar{\phi}_1\left(\vec{X}\right) \right) \frac{\partial F_0}{\partial \mu} \Bigg\}
\end{aligned}$$

Journal of Physics: Conference Series **775** (2016) 012003 doi:10.1088/1742-6596/775/1/012003

Performing the integrals over θ and $\vec{X}$ it is straightforward to show that the generic moment of our gyrocenter distribution function can be reduced to

$$M_{a,b}(\vec{x}) = \frac{2\pi}{m}\int B_0(\vec{x})\left\{\langle F_1(\vec{x}-\vec{r})\rangle + \frac{1}{B_0(\vec{x})}\left[\left(\Omega\frac{\partial F_0}{\partial v_{\shortparallel}} - \frac{q}{c}v_{\shortparallel}\frac{\partial F_0}{\partial \mu}\right)\cdot\left(\vec{A}_1(\vec{x}) - \left\langle\vec{A}_1(\vec{x}-\vec{r})\right\rangle\right) + \right.\right.$$

$$\left.\left. +q\left(\phi_1(\vec{x}) - \langle\bar{\phi}_1(\vec{x}-\vec{r})\rangle\right)\frac{\partial F_0}{\partial \mu}\right]\right\} v_{\shortparallel}^a v_{\perp}^b dv_{\shortparallel} d\mu \qquad (7)$$

Here, the brackets $\langle \ldots \rangle$ denote the θ integrated function, i.e. gyroaveraged quantities. In case of a background Maxwellian distribution function, the generic moment equation can be greatly simplified. Indeed, the terms that multiply the vector potential cancel and the $v_{\shortparallel}$-integration can be performed analytically.

2.4. Field equations

The Poisson equation and the Ampere's law for both the parallel and perpendicular component of the electromagnetic potential can be written in terms of the $M_{0,0}$ and $M_{1,0}$ moments of the perturbed distribution function

$$\nabla_{\perp}^2\phi_1(\vec{x}) = -4\pi\sum_j q_j n_{1,j}(\vec{x}) = -4\pi\sum_j q_j M_{0,0,j}(\vec{x})$$

$$-\nabla_{\perp}^2 A_{1,\shortparallel}(\vec{x}) = \frac{4\pi}{c}\sum_j j_{\shortparallel,1,j}(\vec{x}) = \frac{4\pi}{c}q_j M_{1,0,j}(\vec{x})$$

where we used the following Cartesian coordinate system $\left(\hat{e}_1, \hat{e}_2, \hat{b}_0\right)$.

From Eq.(7) the normalised non-Maxwellian field equations become

$$P\phi_1(\vec{x}) + \mathcal{F}A_{1,\shortparallel}(\vec{x}) = q\pi n_0 B_0\int J_0(\lambda)\, g_1 dv_{\shortparallel} d\mu$$

$$\mathcal{H}A_{1,\shortparallel}(\mathbf{x}) + \mathcal{L}\phi_1(\vec{x}) = qn_0\pi\beta_{ref}\frac{B_0 v_{th}}{2}\int v_{\shortparallel} J_0 g_1 dv_{\shortparallel} d\mu$$

Here, we have introduced the following operators

$$P = k_{\perp}^2\lambda_{De}^2 - \frac{\pi q^2 n_0}{T_0}\int\left(1-J_0^2\right)\frac{\partial F_0}{\partial\mu}dv_{\shortparallel} d\mu \qquad \mathcal{F} = \frac{2\pi q^2 n_0}{m v_{th}}\int\left[\left(1-J_0^2\right)v_{\shortparallel}\frac{\partial F_0}{\partial\mu} - \frac{B_0}{2}\frac{\partial F_0}{\partial v_{\shortparallel}}\right]dv_{\shortparallel} d\mu$$

$$\mathcal{H} = k_{\perp}^2 - \frac{q^2 n_0\pi\beta_{ref}}{m}\int\left[B_0\frac{v_{\shortparallel}}{2}\frac{\partial F_0}{\partial v_{\shortparallel}} - v_{\shortparallel}^2\frac{\partial F_0}{\partial\mu}\left(1-J_0^2\right)\right]dv_{\shortparallel} d\mu$$

$$\mathcal{L} = \frac{q^2 n_0\pi\beta_{ref}}{m v_{th}}\int\frac{\partial F_0}{\partial\mu}\left(1-J_0^2\right)v_{\shortparallel} dv_{\shortparallel} d\mu$$

It is worth noting that for a completely general background distribution function the field equations constitute a system of coupled equations. This system decouples if a Maxwellian distribution function is chosen.

3. Linear simulations

In this paragraph linear electrostatic benchmarks are shown with GS2 and GKW for slowing down fast particles to verify the results shown in the previous section. Furthermore, the relevance of fast particles on the ITG turbulence in a JET hybrid scenario [8] and low magnetic shear ($\hat{s} = 0.5$) has been studied with linear simulations.

Journal of Physics: Conference Series **775** (2016) 012003 doi:10.1088/1742-6596/775/1/012003

3.1. Verification

Linear electrostatic benchmarks have been carried out with the gyrokinetic code GS2 and GKW in order to prove the correctness of the derivation and implementation of the non-Maxwellian Vlasov-Maxwell system of equations. Both these codes have published results from simulations with a slowing-down equilibrium distribution function for a single fast particle species [10]. In this paragraph, growth rates as a function of the fast particle concentration and the alpha particle diffusivity as a function of the electron temperature are shown in Fig.(1) for the same reference case as the one studied in Ref.[10], i.e. cyclone base for a Deuterium, electron and alpha particle plasma. The input parameters are briefly summarised in the following table. As can be seen from Fig.(1) the agreements between codes is excellent. These results prove the validity of both the analytic calculation and the numerical implementation.

$\hat{s}$	q	T_e/T_i	R/L_{T_i}	R/L_{T_e}	R/L_{n_e}	r/a	R/a
0.8	1.4	1.0	6.0	6.0	3.0	0.5	3.0

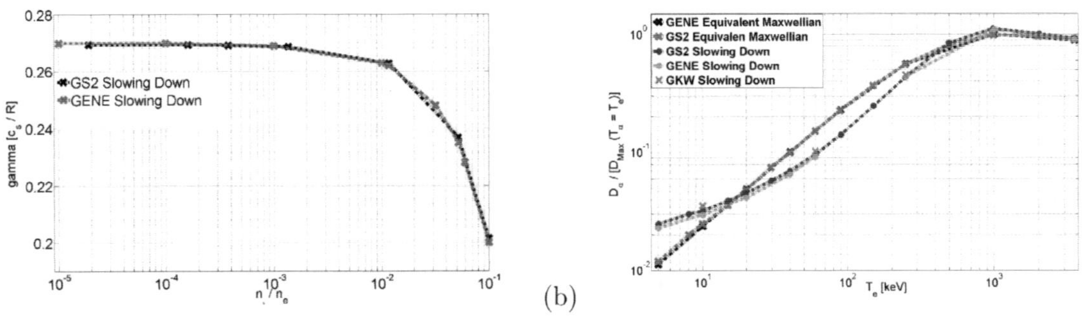

(a) (b)

Figure 1: (a) GENE (red line) and GS2 (black line) growth rates as a function of the α particles density concentration plotted in logarithmic scale. (b) GENE (cyan line), GS2 (blue line) and GKW (magenta cross) α particles diffusivity for slowing down distribution function and GENE (black line), GS2 (red line) α particles diffusivity for equivalent Maxwellian distribution function normalised to the diffusivity of thermal helium plotted in logarithmic scale.

3.2. Microinstability study including more realistic background

In this section a particular JET plasma is revised and analysed with linear simulations [8]. The bulk plasma is composed by Deuterium, electron and Carbon impurities, while the fast particles are fast Deuterium (from NBI) and ^{3}He (from minority heating). The fast ion background distribution function has been modelled with a slowing down, an equivalent Maxwellian and with an asymmetric and anisotropic Maxwellian, in order to study the impact of non Maxwellian background distribution functions on the ITG instability. Low beta electromagnetic effects, collisions, impurities, fast ions, and experimental geometry are included. In the following tables the background plasma and magnetic geometry as well as the fast particles parameters are summarised.

Journal of Physics: Conference Series **775** (2016) 012003 doi:10.1088/1742-6596/775/1/012003

$\hat{s}$	q	T_e/T_i	R/L_{T_i}	R/L_{T_e}	R/L_{n_e}	β_e [%]	ν^*
0.5	1.7	1.0	10	6.8	1.3	0.33	0.038

n_{fD}	n_{3He}	T_{fD}	T_{3He}	$R/L_{T_{fD}}$	$R/L_{T_{3He}}$	R/L_{nfD}	$R/L_{n_{3He}}$
0.06	0.07	9.8	6.9	3	20	13	1.5

3.2.1. Slowing down background In order to study the impact of the fast ion background distribution on microinstabilities in a realistic scenario, we performed linear simulations with five species, modelling the fast particles with a slowing down (defined in Eq.(2)). In Fig.2 linear growth rates are shown for four different cases: equivalent Maxwellian fast particles, slowing down fast particles, slowing down fast Deuterium - Maxwellian ^{3}He and without fast particles (3 species).

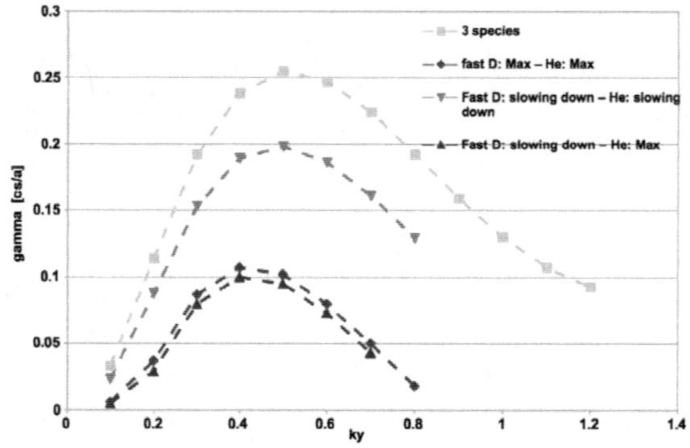

Figure 2: Linear growth rates as function of $k_y\rho_i$ for equivalent Maxwellian fast particles (blue line); Slowing down fast particles (yellow line); Slowing down fast Deuterium - equivalent Maxwellian ^{3}He (green line); without fast particles (red line).

The electromagnetic stabilisation is stronger for the equivalent Maxwellian fast ions simulations, whose growth rates are $\sim 50\%$ lower than the slowing down ones. The most unstable mode is slightly shifted to higher k_y for the slowing down case, matching the k_y related to the most unstable mode in the three species simulation. From the comparison between the slowing down fast Deuterium - equivalent Maxwellian ^{3}He and the both slowing down fast particles growth rates, it is possible to see that the main difference arises when the ^{3}He is modelled with a slowing down distribution. The slowing down fast Deuterium results are slightly more stable than the equivalent Maxwellian ones, while the ^{3}He has an opposite behavior. Slowing down ^{3}He growth rates are much more unstable than the ones carried out using an equivalent Maxwellian. However, a slowing-down distribution function is most likely a poor approximation for the ^{3}He species.

3.2.2. Bi-Maxwellian background In this section we compare the results carried out with the equivalent Maxwellian and the bi-Maxwellian distributions. The latter can be carried out from Eq.(3) in the limit of $T_{\parallel,+} = T_{\parallel,-}$ and is considered as a first order approximation for the ICRH-driven fast ^{3}He [19]. The ^{3}He profiles, i.e. both temperatures and temperature gradients, were extracted from the SPOT/SELFO code [20, 21]. In Fig.3 linear growth rates are shown as a function of $k_y\rho_i$ for five species simulations. The fast Deuterium is modelled with the equivalent Maxwellian distribution function, while for the fast ^{3}He we used the bi-Maxwellian distribution function.

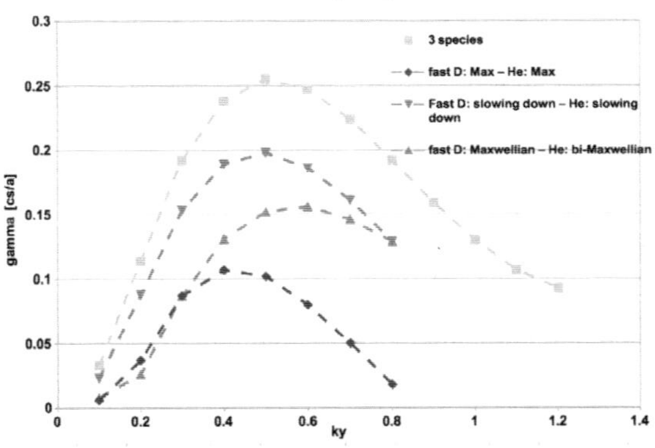

Figure 3: Linear growth rates as function of $k_y\rho_i$ for equivalent Maxwellian fast particles (blue line); Slowing down fast particles (red line); equivalent Maxwellian fast Deuterium - Bi-Maxwellian ^{3}He (green line); without fast particles (yellow line).

As can be seen in Fig.3 the strength of the electromagnetic stabilisation is decreased when modelling the ^{3}He distribution with the bi-Maxwellian. The growth rates are higher than the equivalent Maxwellian ones. In particular, it is worth noting that the difference between the equivalent Maxwellian and bi-Maxwellian growth rates increases with k_y. From Fig.2 and Fig.3 it is possible to observe that the ^{3}He background distribution has a much stronger effect on the plasma mictroturbulence than that of the fast Deuterium.

4. Conclusions

The gyrokinetic turbulence code GENE has been modified and extended in the low-beta electromagnetic case in order to treat any number of non-Maxwellian fast particle species. In particular, slowing down and asymmetric and anisotropic Maxwellian fast particles can now be studied. In order to test and verify the implementation, linear electrostatic benchmarks have been shown for slowing down fusion born alpha particles with existing data from the gyrokinetic codes GS2 and GKW. After these verification studies, a previous equivalent Maxwellian study on a particular JET plasma with fast ion enhanced electromagnetic stabilisation has been revised and analysed with linear simulations. The bulk plasma is composed by Deuterium, electron and Carbon impurities, while the fast particles are NBI fast Deuterium and ICRH ^{3}He. Electromagnetic effects, collisions and experimental geometry have been included. In this context, linear growth rates are studied as a function of the wave number and compared with those obtained using an equivalent Maxwellian. It has been found that also with the more realistic distribution function the increased EM-stabilisation with fast ions still holds, even if the strength of the effect with ICRH is slightly weaker. Furthermore, a lack of sensitivity to the NBI fast ion distribution has been observed. Generally, it has been found that the choice of the ^{3}He background distribution has a stronger impact on the linear results than the fast Deuterium background. These results would be in line with the previous nonlinear findings where gradients higher than the nominal ones had to be employed in order to match the experimental heat fluxes [7, 8]. A corresponding nonlinear analysis will be performed in the near future.

Acknowledgments
The simulations presented in this work were carried out using the TOKP cluster at the Rechenzentrum Garching (RZG), Germany. This work has furthermore been carried out within the framework of the EUROfusion Consortium and has received funding from the Euratom research and training programme 2014-2018 under grant agreement No 633053. The views and opinions expressed herein do not necessarily reflect those of the European Commission.

References
[1] D. Zarzoso et al. 2013 *Phys. Rev. Lett.* **110** 125002
[2] M. Romanelli et al. 2010 *Plasma Physics and Controlled Fusion* **52** 4
[3] J. Garcia et al. 2015 *Nuclear Fusion* **55** 5
[4] G. Tardini et al. 2007 *Nucl. Fusion* **47** 280
[5] C. Holland et al. 2011 *Phys. Plasmas* **18** 056113
[6] C. Bourdelle et al. 2005 *Nucl. Fusion* **45** 110
[7] J. Citrin et al. 2015 *Plasma Physics and Controlled Fusion* **57** 014032
[8] J. Citrin et al. 2013 *Phys. Rev. Lett.* **111** 155001
[9] F. Jenko et al. 2000 *Phys. Plasmas* **7** 1904; See http://genecode.org for code details and access
[10] C. Angioni et al. 2008 *Phys. Plasmas* **15**, 052307
[11] A. G. Peeters et al. 2004 *Phys. Plasmas* **11**, 3748
[12] A. G. Peeters et al. 2009 *Comput. Phys. Commun.* **180**, 2650
[13] M. Kotschenreuther et al. 1995 *Comput. Phys. Commun.* **88**, 128
[14] W. Dorland et al. 2000 *Phys. Rev. Lett.* **85**, 5579
[15] J. Gaffey 1976 *J.Plasma Phyisics* **16** part 2
[16] T. Dannert et al. 2008 *Phys. Plasmas* **15** 062508
[17] Estrada-Mila et al. 2006 *Phys. Plasmas* **13** 112303
[18] A.J. Brizard and T. S. Hahm 2007 *Rev. Modern Physics* **79** 421
[19] R. O. Dendy et al. 1995 *Phys. Plasmas* **2**, 1623
[20] M. Schneider et al. 2011 *Nucl. Fusion* **51**, 063019
[21] J. Hedin et al. 2002 *Nucl. Fusion* **42**, 527

On the relationship between residual zonal flows and bump-on tail saturated instabilities.

X. Garbet[1], P. Donnel[1], C. Ehrlacher[1], E. Caschera[1], R. Dumont[1], M. Faganello[2], V. Grandgirard[1], P. Ghendrih [1], M. Idouakass[2], M. Lesur[3], Y. Sarazin[1], D. Zarzoso[2]

[1]CEA, IRFM, F-13108 St.Paul-lez-Durance cedex, France
[2] Aix-Marseille Université, CNRS, PIIM UMR 7345, Marseille, France
[3] IJL, UMR 7198 CNRS, Université de Lorraine, 54500 Vandoeuvre-lès-Nancy, France

E-mail: xavier.garbet@cea.fr

Abstract. A connection is established between two classical problems: the non linear saturation of a bump-on tail instability in collisionless regime, and the decay of a zonal flow towards a finite amplitude residual. Reasons for this connection are given and commented.

1. Introduction

This paper establishes a connection between two well-known analytical results that are currently used for code verification. The first one is the Rosenbluth-Hinton theory of zonal flow decay [1], which predicts an evolution towards a finite asymptotic amplitude called residual flow. This prediction is routinely used to test and benchmark gyrokinetic codes. The second result addresses the saturated value of the bump-on tail instability in collisionless plasmas. It predicts a bounce frequency of deeply trapped particles that is proportional to the linear growth rate, or equivalently a saturated mode amplitude that goes like the square of the linear drive (see [2] for an overview). This result is an extension of a seminal paper by O'Neil that addresses Landau damping in a nonlinear regime [3] (see also the work by Mazitov [4]). It is used to test codes that solve 2D (x,v) Vlasov/Poisson problems. In fact MHD modes driven by energetic particles often exhibit a quadratic scaling of the mode amplitude with the linear growth rate when the drive is weak, in line with this prediction (cf. [5, 6] and references therein). It turns out that these two results, zonal flow residual and saturated level of bump on tail instability, are connected. The reason for this relationship, which is not entirely fortuitous, is detailed and commented in this paper.

The paper is organized as follows. Section 2 summarizes the O'Neil calculation of Landau damping in non linear regime and its extension to the bump on tail instability. The Rosenbluth-Hinton calculation of the time evolution of zonal flow is reviewed in section 3. Finally the relationship between residual zonal flows and bump-on tail saturated instabilities is discussed in section 4. A conclusion follows.

2. Landau damping in non linear regime - extension to the bump on tail instability

The seminal O'Neil paper [3] aimed at studying Landau damping in collisionless regime in the non linear regime where particle trapping matters. The time evolution of the distribution

Journal of Physics: Conference Series **775** (2016) 012004 doi:10.1088/1742-6596/775/1/012004

function and electric field were calculated. O'Neil realized that the asymptotic state can be captured by calculating the energy transfer from the wave to particles. This calculation was a source of inspiration for many subsequent works (e.g. [7, 8]). The O'Neil work was soon extended to the bump-on tail instability. The saturated state can be calculated by stating that the kinetic energy of resonant particles is fully converted to wave energy. The key point is then to calculate the distribution function. O'Neil prescription consists in using a phase mixing ansatz. The later states that the relaxed distribution function is the average of the unperturbed one along trajectories, due to phase mixing. Hence it is flat within the island, and joins smoothly the unperturbed distribution function outside the separatrix. This prescription is consistent with the detailed calculation of the mode evolution done by O'Neil in the same work. Berk's prescription rather corresponds to a discontinuous distribution function that is flat in the island, and joins the unperturbed distribution right at the separatrix [5]. A third option is to choose a flat distribution function within an annulus, whose width is the island width [2]. Of course other choices are possible as to the contribution of particles near the island separatrix.

2.1. Summary of the O'Neil calculation

Let us start from an arbitrary integrable Hamiltonian system that can be described by a set of angle/action variables $(\mathbf{\Theta}, \mathbf{J})$ such that

$$\begin{aligned} \frac{d\mathbf{\Theta}}{dt} &= \frac{\partial H_{eq}}{\partial \mathbf{J}} = \mathbf{\Omega}_{eq}(\mathbf{J}) \\ \frac{d\mathbf{J}}{dt} &= -\frac{\partial H_{eq}}{\partial \mathbf{\Theta}} = 0 \end{aligned}$$

where $H_{eq}(\mathbf{J})$ is the unperturbed Hamiltonian. This system is perturbed by an oscillation, i.e.

$$H(\mathbf{\Theta}, \mathbf{J}, t) = H_{eq}(\mathbf{J}) - h\cos(\mathbf{n}\cdot\mathbf{\Theta} - \omega t) \tag{1}$$

We look for an asymptotic steady state, i.e. the amplitude h of the perturbed Hamiltonian is a constant (in the original O'Neil paper, the evolution of $h(t)$ was computed). We also introduce the resonance frequency $\Omega(\mathbf{J}) = \mathbf{n}\cdot\mathbf{\Omega}_{eq}(\mathbf{J}) - \omega$. This general formulation allows the treatment of a wide variety of problems. For instance $(\Theta_1, \Theta_2, \Theta_3)$ are the cyclotron, poloidal and toroidal angles for passing particles in a tokamak, while it becomes the cyclotron, bounce and precessional angles for trapped particles. The angle Θ_1 is just $\Theta_1 = kx$ in the O'Neil case of a 1D electrostatic wave, where k is the wave number, and x the position.

The resonant surface is defined by the condition $\Omega(\mathbf{J}) = 0$ in the action space. Near the resonant surface, the actions are of the form $\mathbf{J} = \mathbf{J}_R + \mathbf{n}I$, where $\mathbf{J}_R$ spans the resonant surface, while the action I measures the distance to the resonant surface. The resonant frequency reads $\Omega = CI$, where C is related to the Hessian matrix of the Hamiltonian calculated at the resonant surface

$$C = n_i n_k \left.\frac{\partial^2 H}{\partial J_i \partial J_k}\right|_{\Omega=0} \tag{2}$$

We assume that C is positive without any loss of generality. The trajectory equations read

$$\begin{aligned} \frac{d\xi}{dt} &= \Omega \\ \frac{d\Omega}{dt} &= -\omega_b^2 \sin\xi \end{aligned}$$

where $\xi = \mathbf{n}\cdot\mathbf{\Theta} - \omega t$ is an angle, and $\omega_b = \sqrt{Ch}$ is the bounce frequency of deeply trapped particles. The variable Ω is quite convenient since it is a pulsation and therefore can be compared to other characteristic frequencies, such as the the bounce pulsation ω_b (and a damping rate for dissipative problems). The variable I presents also some advantages as it is an action conjugate to ξ. It is reminded here that the Jacobian is equal to 1 when moving from one to another set of conjugate variables. The new Hamiltonian $\mathcal{H}$ is defined as

$$\mathcal{H} = \frac{1}{2}\Omega^2 - \omega_b^2 \cos\xi \tag{3}$$

$\mathcal{H}$ is a motion invariant, so that the system is still integrable. Its contour lines exhibit the shape of an island (Fig.1). Eqs.(3) show that (ξ, Ω) are conjugate variables with respect to the Hamiltonian $\mathcal{H}$. The set (ξ, Ω) can be completed by a set of conjugate variables $(\mathbf{\Theta}_R, \mathbf{J}_R)$, which span the resonant surface. The phase space volume element reads

$$d^3\mathbf{\Theta} d^3\mathbf{J} = d^2\mathbf{J}_R d^2\mathbf{\Theta}_R dI d\xi \tag{4}$$

To simplify the notations, the explicit dependence on $(\mathbf{\Theta}_R, \mathbf{J}_R)$ will be omitted in the following. In other words, the problem becomes 2D, and decoupled from the dynamics along the resonant surface. In the original O'Neil calculation, which addresses a 1D electrostatic wave, the following correspondences hold

$$h = e\phi \quad ; \quad \Omega = kv - \omega \quad ; \quad \xi = kx - \omega t \quad ; \quad C = \frac{k^2}{m} \quad ; \quad \omega_b = k\left(\frac{e\phi}{m}\right)^{1/2} \tag{5}$$

where v is the velocity, and ϕ the electric potential. We introduce an average ("bounce" average) over time for any function $f(\mathcal{H}, \xi, \epsilon_\Omega)$ where ϵ_Ω is the sign of Ω, which can be written as

$$\langle f \rangle = \frac{1}{\tau} \int_{-\pi}^{\pi} \frac{d\xi}{2\pi} \frac{1}{|\Omega|} f(\mathcal{H}, \xi, \epsilon_\Omega) \tag{6}$$

for passing particles ($\omega_b^2 \le \mathcal{H} \le \infty$), while

$$\langle f \rangle = \frac{1}{2\tau} \int_{-\xi_0}^{\xi_0} \frac{d\xi}{2\pi} \frac{1}{|\Omega|} \left(f(\mathcal{H}, \xi, 1) + f(\mathcal{H}, \xi, -1) \right) \tag{7}$$

for trapped particles ($-\omega_b^2 \le \mathcal{H} \le \omega_b^2$). Here Ω is to be understood as a function of $(\mathcal{H}, \xi, \epsilon_\Omega)$

$$\Omega = \epsilon_\Omega \sqrt{2} \left(\mathcal{H} + \omega_b^2 \cos\xi\right)^{1/2} \tag{8}$$

The bounce angle ξ_0 is the positive root of $\omega_b^2 \cos\xi_0 = \mathcal{H}$. The normalizing time τ is defined as

$$\tau = \int_{-\xi_0}^{\xi_0} \frac{d\xi}{2\pi} \frac{1}{|\Omega|} \tag{9}$$

with $\xi_0 = \pi$ for passing particles. Note that τ is a continuous function of $\mathcal{H}$, and differs from the bounce period T_b. More precisely $\tau = \frac{T_b}{2\pi}$ for passing particles and $\tau = \frac{T_b}{4\pi}$ for trapped particles.

2.2. *Saturation level*

A subtle point is that the kinetic energy should be calculated in the laboratory frame, while the calculation of the distribution function is done in the wave frame. The Hamiltonian in the laboratory frame $\mathcal{H}_{lab}$ is related to the one in the wave frame $\mathcal{H}$ via the relation $\mathcal{H} = \mathcal{H}_{lab} - \omega I$, or equivalently $\mathcal{H}_{lab} = \mathcal{H} + \Omega\frac{\omega}{C}$. The Hamiltonian $\mathcal{H}$ does not participate in the resonant transfer

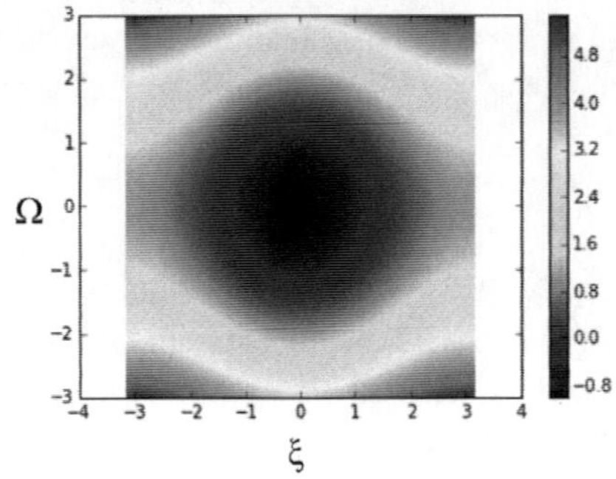

Figure 1. *Island shape ($\omega_b = 1$).*

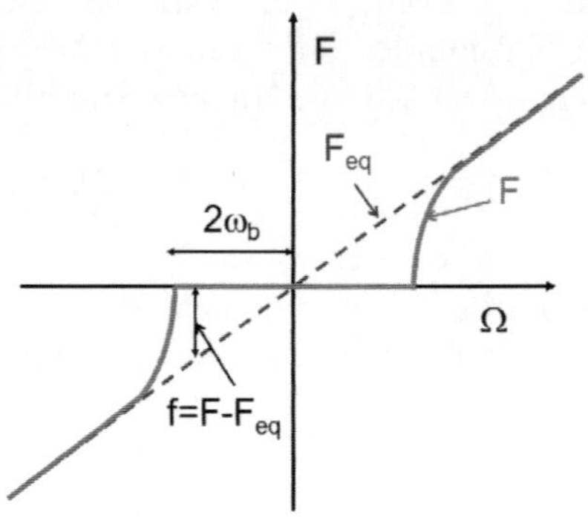

Figure 2. *Schematic view of the distribution function at the island O point*

of energy due to parity reason. Indeed $f = F - F_{eq}$, where F_{eq} is the unperturbed distribution function, is an odd function of Ω. Therefore the resonant transfer of kinetic energy is

$$\Delta E = \frac{\omega}{C}\int_{-\infty}^{+\infty} dI \int_{-\pi}^{+\pi} \frac{d\xi}{2\pi} f\Omega \tag{10}$$

Moreover F_{eq} which can be expanded near the resonant surface $\Omega = 0$, i.e. $F_{eq} = F_0 + F_0'\Omega$. The phase mixing prescription states that the relaxed distribution function is the average of the initial one over trajectories, i.e.

$$f = F - F_{eq} = F_0'\left[\langle\Omega\rangle - \Omega\right] \tag{11}$$

Note that $\langle\Omega\rangle$ is zero for trapped particles, since Ω changes sign when the particle moves back and forth due to trapping. A schematic view of the distribution function at the O point of the island is shown in Fig.(2). A change of variables yields the following result

$$\Delta E = 2\frac{\omega}{C^2}F_0'\int_{-\omega_b^2}^{+\infty} d\mathcal{H}\mathcal{T}\left\{\langle\Omega\rangle^2 - \left\langle\Omega^2\right\rangle\right\} \tag{12}$$

It appears that most of the energy given by particles to the mode comes from trapped particles, and to a lesser extent from barely passing particles. When balancing the energy transfer with the mode energy density $\Lambda_\omega h^2$, one gets a prediction for the field amplitude $\Delta E + \Lambda_\omega h^2 = 0$. In the O'Neil original problem (Landau damping of a 1D electrostatic wave) , the energy density is just $\Lambda_\omega h^2 = \frac{\epsilon_0}{2}\mathcal{E}^2$, where $\mathcal{E}$ is the electric field amplitude and ϵ_0 is the vacuum permittivity, so that $\Lambda_\omega = \epsilon_0 k^2/2e^2$. For other problems, the expression of Λ_ω can be more intricate, but it remains generically independent of the perturbed Hamiltonian h. The final result can be formulated as

$$\omega_b = \mathcal{K}_{BT}\gamma_L \tag{13}$$

where $\gamma_L = \frac{\pi}{4}\omega\frac{F_0'}{\Lambda_\omega}$ is the linear growth rate. The constant $\mathcal{K}_{BT}$ is given by the following relation

$$\mathcal{K}_{BT} = \frac{32}{\pi}\int_0^\infty \frac{d\kappa}{\kappa^3}\frac{\mathcal{T}}{\omega_b}\left\{\left\langle\Omega^2\right\rangle - \langle\Omega\rangle^2\right\} \tag{14}$$

where

$$\Omega = \frac{2\omega_b}{\kappa}\epsilon_\Omega\left[1 - \kappa^2\sin^2\left(\frac{\xi}{2}\right)\right]^{1/2} \tag{15}$$

and κ is the the trapping parameter defined as $\kappa^2 = \frac{2\omega_b^2}{\mathcal{H}+\omega_b^2}$. Trapping corresponds to $1 \leq \kappa^2 \leq \infty$ while passing particles are characterized by $0 \leq \kappa^2 \leq 1$.

3. Zonal flow residual

3.1. Formal expression

In 1998 Rosenbluth and Hinton calculated the time evolution of a zonal flow and predicted the asymptotic value, called zonal flow residual [1]. The calculation consists in initializing the electric potential with a perturbation of the form $\phi(\mathbf{x}, t = 0) = \phi_0 e^{ik\psi}$, where k is the radial wavenumber, and ψ is minus the poloidal magnetic flux normalized to 2π, also equal to $-A_\varphi R$, where A_φ is the toroidal component of the vector potential, and R the major radius. It is a function of the radial coordinate, labeled r. The evolution of $\phi(\mathbf{x}, t)$ is calculated by solving the gyrokinetic Vlasov equation coupled to the electroneutrality equation. The ratio of the potential residual to its initial value is

$$\frac{\phi_\infty}{\phi_0} = \frac{1 - \int d^3\mathbf{p} F_{eq} J_0^2}{1 - \int d^3\mathbf{p} F_{eq} J_0^2 J_b^2} \tag{16}$$

where J_0 and J_b are related to averages over the cyclotron and drift motions, F_{eq} is the unperturbed distribution function. These are scalars for Fourier modes.

3.2. Field shielding

The gyroaverage function is the usual one, and is approximated by the following form in the large scale limit $J_0^2 \simeq 1 - \frac{1}{2}k_\perp^2 \rho_c^2$, where ρ_c is the kinetic ion Larmor radius. The guiding center radial coordinate reads $\psi = \psi_0 + \hat{\psi}$, where ψ_0 is the time average position and $\hat{\psi}$ describes the shift with respect to the reference magnetic surface $\psi = \psi_0$. For a periodic potential, the transit average operator is defined as

$$J_b = \left\langle e^{ik\hat{\psi}} \right\rangle \simeq 1 - k^2 \left\langle \hat{\psi}^2 \right\rangle \tag{17}$$

where the bracket is an average over the drift motion. At first order in $k^2\rho_c^2$, the shielding term reads

$$1 - \int d^3\mathbf{p} F_{eq} J_0^2 J_b^2 \simeq k^2 \left\langle \frac{1}{2}\rho_c^2 + \int d^3\mathbf{p} F_{eq} \hat{\psi}^2 \right\rangle = k^2 \left(\rho_i^2 + \delta^2\right) \tag{18}$$

where $\rho_i = \frac{mT}{eB}$ is the thermal ion Larmor radius, and

$$\left(\frac{d\psi}{dr}\right)^2 \delta^2 = \int d^3\mathbf{p} F_{eq} \left\langle \hat{\psi}^2 \right\rangle \tag{19}$$

is the radial size of the drift orbit squared. The expression for the radial displacement $\hat{\psi}$ is obtained by using the conservation of the canonical toroidal kinetic momentum $P_\varphi = m\frac{I}{B}v_\parallel - e\psi$, namely

$$\hat{\psi} = \frac{m}{e} I \left(\frac{v_\parallel}{B} - \left\langle \frac{v_\parallel}{B} \right\rangle_b \right) \tag{20}$$

where the index 'b' indicates a bounce average for particles trapped in the magnetic field due to the mirror force, while it labels a transit time average for passing particles. Here $I = B_T R$ is the product of the toroidal field by the major radius, B is the modulus of the magnetic field, $v_\parallel$ is the particle velocity along the magnetic field, m and e are the ion mass and charge. This gives an expression of the width δ

$$\left(\frac{d\psi}{dr}\right)^2 \delta^2 = \frac{1}{2}\left(\frac{mI}{e}\right)^2 \int d^3\mathbf{p} F_{eq} \left[\left\langle \left(\frac{v_\parallel}{B}\right)^2 \right\rangle_b - \left\langle \frac{v_\parallel}{B} \right\rangle_b^2 \right] \tag{21}$$

Journal of Physics: Conference Series **775** (2016) 012004 doi:10.1088/1742-6596/775/1/012004

This is the main result derived by Rosenbluth and Hinton in arbitrary geometry. We now restrict the analysis to circular concentric magnetic surfaces with large aspect ratio, so that r is now the minor radius of a magnetic surface, and θ is the poloidal angle. The hamiltonian reads

$$H = \frac{1}{2}mv_\parallel^2 + \mu B_0(1 - \epsilon\cos\theta) \tag{22}$$

where $\epsilon = \frac{r}{R_0}$ is the inverse aspect ratio. This Hamiltonian is obviously identical to the one addressed in the previous section, with the following correspondences

$$h = \mu B_0\epsilon \;\; ; \;\; \Omega = \frac{v_\parallel}{qR_0} \;\; ; \;\; \xi = \theta \;\; ; \;\; C = \frac{1}{mq^2R_0^2} \;\; ; \;\; \omega_b = \frac{1}{qR_0}\left(\frac{\mu B_0\epsilon}{m}\right)^{1/2} \tag{23}$$

and $\mathcal{H} = C(H - \mu B_0)$. Hence the problem is related again to particle trapping. However the reader should be warned that trapping here is due to a magnetic mirror force, not to the mode itself. The amplitude of the latter is very small, so that the problem can be treated linearly with respect to ϕ. Moreover, it can be easily verified that for a Maxwellian distribution function

$$F_{eq}d^3\mathbf{p} = (2\epsilon)^{1/2}\frac{2}{\sqrt{\pi}}dEE^{1/2}e^{-E}\frac{2d\kappa}{\left[(1-\epsilon)\kappa^2 + 2\epsilon\right]^{3/2}}\omega_b\tau \tag{24}$$

where E is the energy normalized to the temperature. In the limit of large aspect ratio $\epsilon \to 0$ and precluding that the largest contribution comes from the separatrix neighborhood $\kappa \simeq 1$, a compact expression of the zonal flow residual Eq.(16) is found

$$\frac{\phi_\infty}{\phi_0} = \frac{1}{1 + \mathcal{C}_{RH}\frac{q^2}{\sqrt{\epsilon}}} \tag{25}$$

where

$$\mathcal{C}_{RH} = 3\sqrt{2}\int_0^\infty \frac{d\kappa}{\kappa^3}\frac{\tau}{\omega_b}\left\{\left\langle\Omega^2\right\rangle - \langle\Omega\rangle^2\right\} \tag{26}$$

4. Relationship between residual zonal flows and bump-on tail saturated instabilities

4.1. Numerical values and metrology

The two constants $\mathcal{C}_{RH}$ and $\mathcal{K}_{BT}$ are related to the quantity $\mathcal{I}$

$$\mathcal{I} = \int_0^\infty \frac{d\kappa}{\kappa^3}\frac{\tau}{\omega_b}\left\{\left\langle\Omega^2\right\rangle - \langle\Omega\rangle^2\right\} \tag{27}$$

through the relations $\mathcal{K}_{BT} = \frac{32}{\pi}\mathcal{I}$ and $\mathcal{C}_{RH} = 3\sqrt{2}\mathcal{I}$. A straightforward calculation shows that

$$\begin{aligned} \mathcal{I} &= 2\int_0^1 \frac{d\kappa}{\kappa^4}\left[\frac{2}{\pi}E\left(\kappa^2\right) - \frac{\pi}{2}\frac{1}{K\left(\kappa^2\right)}\right] \\ &+ 2\int_1^{+\infty}\frac{d\kappa}{\kappa^3}\left[\left(\frac{1}{\kappa^2} - 1\right)\frac{2}{\pi}K\left(\frac{1}{\kappa^2}\right) + \frac{2}{\pi}E\left(\frac{1}{\kappa^2}\right)\right] \end{aligned} \tag{28}$$

where K and E are complete elliptical functions of first and second kinds. A numerical calculation yields $\mathcal{I} \simeq 0.38$ [12], which implies $\mathcal{C}_{RH} \simeq 1.61$ and $\mathcal{K}_{BT} \simeq 3.84$. The value $\mathcal{C}_{RH} \simeq 1.6$ has been recovered in a number of gyrokinetic codes, and is now considered as a "must do" verification test in gyrokinetic computation. However less success was met for the bump-on tail instability since numerical simulations rather find $\mathcal{K}_{BT} \simeq 3.2$ [9, 11]. An alternative recipe consists in postulating that trapped particles only contribute to the mode saturation. It turns out that $\mathcal{I}$ is analytic in that specific case, namely $\mathcal{I} = \frac{8}{9\pi}$, and $\mathcal{K}_{BT} = \frac{256}{9\pi^2} \simeq 2.9$ [5]. Another calculation based on O'Neil prescription yields $\mathcal{K}_{BT} \simeq 3.5$ [10]. Finally a radical option is to choose a flat distribution function within an annulus, whose width is the island width. This procedure is fully analytic and leads to $\mathcal{K}_{BT} = \frac{32\sqrt{2}}{3\pi} \simeq 4.8$ [2]. All these results differ from the numerical findings. A value that is too low is found when minimizing the role of passing particles, while an overweight of passing particles leads to an estimate that is too large. So it appears that only a fraction of particles near the separatrix participate to the energy exchange. In particular it looks like barely passing particles do not contribute as expected. This is not surprising since the bounce time is very large near the separatrix (and infinite right on it), so that a mixing argument is questionable for particles in this region of the phase space. This is consistent with the analysis of the contributions of particles to the Rosenbluth-Hinton constant $\mathcal{C}_{RH}$ (and therefore of $\mathcal{K}_{BT}$). It appears that barely trapped particles provide the largest contribution (see Fig. 3). This structure in the velocity space was verified numerically by gyrokinetic codes [13]. Another important test is the dynamics of the mode evolution, which is also analytical for both problems, and quite different. Time evolution has been verified numerically with success in both cases. As a final note, we stress that a steady solution is not unique by far for bump-on tail instabilities, in particular when dissipation is accounted for. Quite often the system evolves to unsteady solutions, which can be periodic, or chaotic, thus leading to frequency chirping [14, 15, 11, 6]. These important developments are beyond the scope of the present paper.

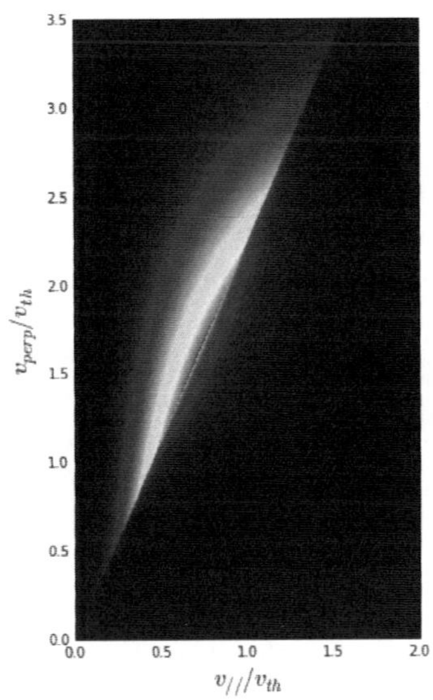

Figure 3. Contour lines of the integrand of $\mathcal{I}$ in the $v_\parallel$,$v_\perp$ space at $\theta = 0$.

4.2. Pure coincidence or some common physics?

The previous sections show that a relationship connects the bump-on tail constant with the Rosenbluth-Hinton constant, i.e. $\mathcal{K}_{BT} = \frac{16\sqrt{2}}{3\pi}\mathcal{C}_{RH}$. This is rather unexpected as the two results correspond to different physics. On the one hand, the bump-on tail saturation comes from a flattening of the distribution function near the resonance surface, leading to a weakening of the drive until saturation. On the other hand, the zonal flow residual comes from a shielding effect of the initial perturbation due to the particle response to the field. So the fast conclusion is that this relationship is fortuitous. There are nevertheless common features : particle trapping, absence of collisions, and more importantly momentum conservation. The relation Eq.(10) shows that the energy transfer in the bump-on-tail instability per unit time can be related to an exchange of momentum between the wave and particles. The power is the drag force $F = \frac{dP}{dt}$ times the wave phase velocity $\frac{\omega}{k}$. Here P is the momentum exchange between the wave and resonant particles

$$P = k \int_{-\infty}^{+\infty} dI \int_{-\pi}^{+\pi} \frac{d\xi}{2\pi} I f = m^2 \int_0^{\lambda} \frac{dx}{\lambda} \int_{-\infty}^{+\infty} dv \left(v - \frac{\omega}{k}\right) f \tag{29}$$

where $\lambda = \frac{2\pi}{k}$ is the wavelength. The zonal flow residual is also related to momentum conservation, though in a different way. As mentioned above, the shielding effect comes from the particle displacement due to drift motion. The poloidal variation of the magnetic field in a tokamak can be seen as a static wave, that exchanges momentum with particles. The change of parallel velocity induces a radial displacement, which results from the conservation of toroidal canonical momentum. This establishes a link between the particle radial shift and the parallel velocity Eq.(20). The displacement is directly proportional to the variance of velocity, hence the $\mathcal{I}$ integral. In the case of the bump-on tail instability, the variance comes in due to the structure of the distribution function. This is probably the reason why the analytic calculation works better for the Rosenbluth-Hinton shielding effect than for the bump-on tail mode saturation: the distribution function results from a complex dynamics, and the assumption of phase mixing is easily broken, in particular near the separatrix.

5. Conclusion

In summary it is found that the saturation of a bump-on tail instability in collisionless regime, and the zonal-flow residual are connected. More precisely, the two constants that appear in these classical results, namely the ratio of the trapping frequency to the linear growth rate, and the zonal flow residual normalized to its initial value, are connected via a simple formula. From the numerical point of view, the prediction for the zonal flow residual works better than the one for the bump-on tail instability. This indicates that for zonal flow decay, all particles play their expected role, while for the bump-on tail problem, particles near the separatrix do not behave as foreseen. This means that the distribution function is flat within the island, as predicted, but exhibits a transition near the separatrix sharper than expected on the basis of phase mixing argument. The relationship between the two problems seems at first sight fortuitous, since the physics that is involved is different. However particle trapping and momentum conservation appear as common denominators, thus explaining partially the link between these expressions.

Acknowledgements

Drs H. Berk and A. Smolyakov are warmly acknowledged for fruitful discussions. This work has also taken benefit from discussions within the NLED Eurofusion Enabling Research project.

References

[1] Rosenbluth M N and Hinton F L 1998 *Phys. Rev. Lett.* **80** 724
[2] Galeev A A and Sagdeev R S 1973 in Review of Plasma Physics vol.7, p.1 "Nonlinear Plasma Theory" [english version: 1979 Consultant Bureau, New York]
[3] O'Neil T 1965 *Phys Fluids* **8** 2255
[4] Mazitov R K 1965 *Zhurnal prikladnoi mekhlaniki i tekhnicheskoi fiziki* **1** 27
[5] Berk H L, Breizman B N, Fitzpatrick J , Pekker M S, Wong H V, and Wong K L, Physics of Plasmas 1996 **3** 1827
[6] Liu Chen and Zonca F 2016 Rev. Mod. Phys. **88** 015008
[7] Isichenko M B *Phys. Rev. Lett.* **78** 2369 (1997)
[8] Manfredi G *Phys. Rev. Lett.* **79** 2815 (1997)
[9] Fried B D, Liu C S, Means R W, and Sagdeev R Z, National Technical Information Service Document No. AD730123 PGG-93, University of California, Los Angeles 1971. Copies may be ordered from the National Technical Information Service, Springfield, Virginia 22161.

[10] Levin M B, Lyubarskii M G, Onishchenko I N, Shapiro V D, and Shevchenko V I 1972 *Zh. Eksp. Teor. Fiz.* **62** 1725 [1972 *Sov. Phys. JETP* **35** 898]
[11] Lesur M, Idomura Y, and Garbet X 2009 *Phys. Plasmas* **16** 092305
[12] Rosenbluth M N et al 1996 *Nucl. Fusion* **36** 55 . See Eq.(50) and paragraph thereafter.
[13] Watanabe T-H and Sugama H Nucl. *Fusion 2006* **46** 24
[14] Berk H L, Breizman B N and Petviashvili N V 1997 *Phys. Lett. A* **234** 213
[15] Vann R G L et al. 2003 *Phys. Plasmas* **10** 623

Cross-polarization scattering of diffracting electron-cyclotron beams in a turbulent plasma with the WKBeam code

L. Guidi[1,2], O. Maj[1], H. Weber[1], A. Köhn[1], A. Snicker[1] and E. Poli[1]

[1]Max Planck Institute for Plasma Physics, Boltzmannstr. 2, 85748 Garching, Germany

[2]Technische Universität München, Numerical Methods for Plasmaphysics (M16), Boltzmannstr. 3, 85748 Garching, Germany

E-mail: `lorenzo.guidi@ipp.mpg.de`

Abstract. In turbulent plasmas, density fluctuations are expected to scatter radiofrequency wave beams, causing a degradation of the beam quality and thus reducing their performance. The WKBeam code is a Monte-Carlo solver for the wave kinetic equation, which describes such an effect, so far limited to a single wave mode so that it cannot account for cross-polarization scattering. In this work a new feature of the WKBeam code is presented, which allows the analysis of cross-polarization scattering of electron-cyclotron (EC) wave beams in realistic tokamak scenarios.
We prove the convergence of the numerical scheme and give a preliminary overview of such effects in ITER scenarios.

1. Introduction

It has been suggested that in next-generation tokamak machines such as ITER, scattering by turbulent fluctuations can deteriorate the quality of electron-cyclotron (EC) beams and thus impair applications relying on high localization of the power deposition [1, 2, 3]. The WKBeam code [4] provides numerical simulations for the propagation of EC waves in a magnetically confined plasma, taking into account scattering phenomena due to density fluctuations. The code is based on a Monte-Carlo solution of the wave kinetic equation in the geometrical optics phase-space [5] for a single wave mode. Relevant quantities - e.g., the power deposition profile - can be analyzed in realistic tokamak scenarios for either the ordinary (O) or extraordinary (X) mode. The method can also treat diffracting beams and short-scale fluctuations, essentially being limited only by the applicability of the Born scattering approximation, which imposes an upper bound on the density spatial correlation function.

Under certain circumstances, however, plasma turbulence can induce mode-to-mode transitions, i.e., part of the injected power, initially assigned to the desired wave mode, can be transferred to another mode with different polarization. Cross-polarization scattering could be significant, in particular, in presence of large amplitude turbulent fluctuations in the tokamak edge, where the cold-plasma wave modes are nearly degenerate and the turbulence spectrum might induce a "jump" in wavevector space between the two branches. The aim of this work is to design and implement a numerical scheme for WKBeam in order to account for such cross-polarization scattering events. Specifically, we design a Monte-Carlo algorithm for the wave kinetic equation (WKE) in the constrained steady-state form which is relevant to time-harmonic coherent wave fields, namely,

$$\{H_\alpha, w_\alpha\} = -2\gamma_\alpha w_\alpha + \mathcal{S}_\alpha(\{w_\beta\}), \tag{1}$$

$$H_\alpha w_\alpha = 0, \tag{2}$$

$$w_\alpha|_\Sigma = w_\alpha^0, \tag{3}$$

where the unknown is the statistically averaged *Wigner function* $w_\alpha = w_\alpha(x, N)$ representing the wave energy density carried by the mode α ($\alpha = X$ and $\alpha = O$ for the extra-ordinary and ordinary mode, respectively). More specifically, w_α is a (generalized) function defined on the standard geometrical optics phase-space with coordinates (x, N), where $x = (x^1, \ldots, x^d)$ is the position in the physical space normalized to the scale length L of the plasma equilibrium density profile, and $N = (N_1, \ldots, N_d)$ is the refractive index vector ($N = ck/\omega$, where c is the speed of light in the phase-space and ω the angular frequency of the wave beam). Eq. (1) describes the transport of the Wigner function in the phase-space. Particularly, $\{H_\alpha, w_\alpha\} = \nabla_N H_\alpha \cdot \nabla_x w_\alpha - \nabla_x H_\alpha \cdot \nabla_N w_\alpha$ are the Poisson brackets of Hamiltonian ray theory, γ_α is the absorption coefficient and, $S_\alpha(\{w_\beta\})$ accounts for scattering due to density fluctuations. Specifically,

$$\mathcal{S}_\alpha(\{w_\beta\})(x, N) = \sum_\beta \int_{\mathbb{R}^d} \sigma_{\beta\alpha}(x, N', N) w_\beta(x, N') dN' - \Sigma_\alpha(x, N) w_\alpha(x, N), \tag{4}$$

where $\Sigma_\alpha(x, N) = \sum_\beta \int_{\mathbb{R}^d} \sigma_{\alpha\beta}(x, N, N')dN'$. The scattering cross-section $\sigma_{\alpha\beta}(x, N, N')$ is related to the probability of the transition $(\alpha; x, N) \to (\beta; x, N')$, i.e., a ray in the mode α at (x, N) being scattered to a ray of mode β and refractive index N' ($\alpha = \beta$ means scattering to the same mode); the position x never changes. Eq. (2) plays the role of a dispersion relation, while Eq. (3) gives a boundary condition on the surface Σ, which represents the launching mirror in the EC system. A statistical description of turbulence combined with tools from the calculus of pseudodifferential operators provides the theoretical framework for the derivation of the WKE from the relevant wave equation for the electric field of the beam, accounting for density fluctuations and weak absorption, in the short-wavelength limit [6, 5].

2. Description of the algorithm

The code WKBeam [4] provides a numerical solution of the WKE in realistic tokamak scenarios, together with a set of post-processing tools which allow an analysis of the quantities of interest for fusion research. Eqs. (1)-(3) are not solved directly but the solution $w_\alpha(x, N)$ is recovered from that of an auxiliary Cauchy problem, namely,

$$\begin{aligned}\frac{\partial u_\alpha}{\partial t} + \{H_\alpha, u_\alpha\} &= -2\gamma_\alpha u_\alpha + \mathcal{S}_\alpha(\{u_\beta\}), \\ u_\alpha|_{t=0} &= u_\alpha^0,\end{aligned} \tag{5}$$

with initial condition u_α^0 chosen so that $w_\alpha = \int u_\alpha dt$ satisfies (1)-(3). The path followed is very similar to standard ray tracing, coupled with a random generation of scattering events. The beam is modeled as an ensemble of markers in the phase space which follow the rays and randomly modify their trajectory in presence of the fluctuations. The code as presented in [7] was able to solve the problem for a single wave mode (O-mode or X-mode), neglecting the terms $\sigma_{\alpha\beta}$ with $\alpha \neq \beta$ in (4). In this work a modification of the algorithm is presented, such that those terms are actually taken into account. With the new algorithm, the possibility of energy transfer to undesired polarization modes can be investigated. Following the theory of Monte-Carlo methods for transport processes [8], we design a numerical scheme which approximates a stochastic process $\{(X(t), \mathcal{N}(t), \alpha(t))\}$, constructed so that its probability density solves the Cauchy problem (5) with $\gamma_\alpha = 0$. The parameter $t \in [0, T]$ along each stochastic ray trajectory is discretized with constant step $\Delta t = T/n_s$, thus defining the intervals $I_j = [t_j, t_{j+1})$, $t_j = j\Delta t$, $j = 0, \ldots, n_s - 1$. We denote by

$$(x_j^\pm, N_j^\pm) = \lim_{t \to t_j^\pm} (X(t), \mathcal{N}(t))$$

the left and right limits of the stochastic ray path at each node $t = t_j$. Given a number N_m of markers labeled by the index $i \in \{0, \ldots, N_m - 1\}$, the algorithm consists of the following operations:

(i) Starting points $\{(X_i(0), \mathcal{N}_i(0), \alpha_i(0))\} \in \mathbb{R}^d \times \mathbb{R}^d \times \{O, X\}$ are sampled from the initial condition $\{u_\alpha^0\}_{\alpha=O,X}$ of the problem.

(ii) In each interval I_j, $\alpha_i(t) = \alpha_{i,j}$ is constant, and the trajectories $(X_i(t), \mathcal{N}_i(t))$ follow the Hamiltonian field of H_α, with$\alpha = \alpha_{i,j}$; we compute the limits $x_{i,j+1}^-$, $N_{i,j+1}^-$.

(iii) The position does not change, i.e., $x_{i,j+1}^+ = x_{i,j+1}^-$.

(iv) The number of scattering events in the interval is extracted from a Poisson distribution of parameter $\bar{\lambda}\Delta t$, with

$$\bar{\lambda} = \max_{\alpha,x,N} \lambda_\alpha(x, N) := \max_{\alpha,x,N} \sum_\beta \int_{\mathbb{R}^d} \sigma_{\alpha\beta}(x, N, N')dN'.$$

(v) For each of such events, scattering from mode α to mode β occurs with probability

$$\tilde{p}_{\alpha\beta} = \frac{\Sigma_{\alpha\beta}}{\lambda_\alpha} := \frac{\int_{\mathbb{R}^d} \sigma_{\alpha\beta}(x, N, N')dN'}{\lambda_\alpha}, \qquad (x, N) = (x_{i,j+1}^-, N_{i,j+1}^-).$$

The target mode $\alpha_{i,j+1}$ is selected accordingly and the new value $N_{i,j+1}^+$ is determined through the Metropolis-Hastings algorithm from the probability density

$$\tilde{\rho}_{\alpha\beta} = \frac{\sigma_{\alpha\beta}}{\lambda_\alpha}, \qquad \alpha = \alpha_{i,j+1}, \quad \beta = \alpha_{i,j}. \tag{6}$$

Journal of Physics: Conference Series **775** (2016) 012005 doi:10.1088/1742-6596/775/1/012005

(vi) Iterate to the next interval I_{j+1}, with initial condition $(x^+_{i,j+1}, N^+_{i,j+1}, \alpha_{i,j+1})$.

By standard techniques [8], one can prove that the probability density of such process approximates the density of $\{X(t), \mathcal{N}(t), \alpha(t)\}$ which solves the auxiliary Cauchy problem (5), with $\gamma_\alpha = 0$. Absorption is then introduced by weighting each marker with the weight $\exp\left(-2\int_0^t \gamma_\alpha dt'\right)$.

3. Verification and convergence

In order to test the algorithm and its convergence, we consider a stand-alone test problem, in which the particular form of the scattering operator allows the use of a pseudospectral solver to compute a reference solution. The simplified model consists in considering an equivalent Cauchy problem (5) in a 2D phase-space (one coordinate x in space, one coordinate y in momentum), with the model cross-section

$$\sigma_{\alpha\beta}(x, y, y') = \sigma_{\alpha\beta}(x, y - y') = s_{\alpha\beta} e^{-L_c^2 (y-y')^2}, \quad \alpha = 0, 1, \tag{7}$$

and Hamiltonians with associated Hamiltonian fields $X_\alpha = (\partial_y H_\alpha, -\partial_x H_\alpha)$ given by

$$X_0 = (-1, -1/2), \quad X_1 = (-1, 1/2).$$

The numbers $s_{\alpha\beta} \in \mathbb{R}_+$ quantify the scattering from mode α to mode β, and L_c plays the role of the turbulence correlation length normalized to the scale length L; the spectral width in (7) is $\propto 1/L_c$. The absorption term, which does not take parte to the definition of the relevant stochastic process, is omitted in this test-model. Under these assumptions, the integral operator $\mathcal{S}_\alpha$ amounts to a convolution, making the pseudospectral solver a suitable tool to build a reference deterministic solution.

Fig. 1 shows a test run with an initial Gaussian distribution in the mode $\alpha = 0$. Panels 1a and 1b show the pseudospectral solution, while panels 1c and 1d the Monte-Carlo solution reconstructed by means of a kernel density estimation (KDE). Each panel shows the reconstructed solution $u_\alpha(t, x, y)$ at various values of the parameter $t = 0, 1, \ldots, 5$. The contours of the pseudospectral solution show excellent qualitative agreement with those of the reconstructed Monte-Carlo distribution. The solution evolves from the initial position, set at $(x, y) = (5.5, 3.4)$, towards the left-hand side and scatters in momentum y, into both the launched mode $\alpha = 0$ (panels 1a/1b) and the secondary mode $\alpha = 1$ (panels 1c/1d), which is not present at $t = 0$. The streamlines of the two Hamiltonian fields X_α for the corresponding mode are represented (black arrows). One can notice a deflection of the center of mass of the solution from the streamline trajectory. This effect is entirely due to cross-polarization scattering: energy is transferred to the "passive" mode $\alpha = 1$, flows along the streamlines of X_1 and has a finite probability of being scattered back into $\alpha = 0$, effectively producing a mechanism to transfer energy perpendicularly to the Hamiltonian field.

Fig. 2 shows the convergence of the algorithm in the standard Monte-Carlo sense. The error on the energy density at a fixed time,

$$\epsilon_t := \max_K |\mu_{N_m}(K) - \nu(K)|,$$

Journal of Physics: Conference Series **775** (2016) 012005 doi:10.1088/1742-6596/775/1/012005

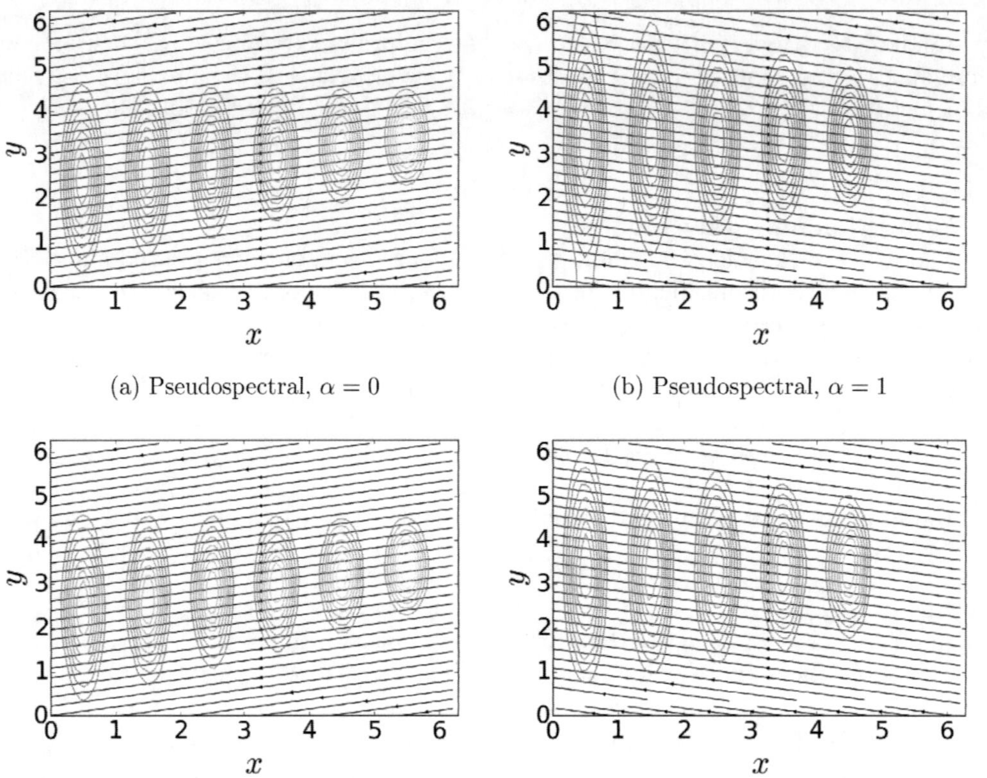

(a) Pseudospectral, $\alpha = 0$

(b) Pseudospectral, $\alpha = 1$

(c) Monte-Carlo, $\alpha = 0$

(d) Monte-Carlo, $\alpha = 1$

Figure 1: Comparison between the Monte-Carlo scheme and a pseudospectral solver. The contours correspond to the levels $u_\alpha = a \cdot \max_{x,y} u_\alpha(t, x, y)$, with $a \in \{0.3, 0.4, \ldots, 0.9\}$.

goes to zero as $1/\sqrt{N_m}$. Here the max is computed for K running over the cells of the dual grid of the pseudospectral solution, μ_{N_m} is the measure that counts the number of markers in a certain set, and ν is the measure associated to the pseudospectral solution.

4. Preliminary physics results

We present here some examples of simulations performed with WKBeam in realistic tokamak scenarios, in order to provide a taste of what kind of situations could be investigated thanks to the cross-polarization scattering algorithm. A crucial role in this framework is played by the model used for the density fluctuations. In all cases the two-point density correlation function - which largely determines the cross section $\sigma_{\alpha\beta}$ - is taken in the form

$$\mathbb{E}\left(\delta n_e(x)\delta n_e(x')\right) = C\left(\frac{x + x'}{2}, x - x'\right) = n_e^2(\rho)F^2(\rho, \theta)\mathrm{e}^{-\frac{1}{2}(x-x')^T A(x-x')}, \quad (8)$$

Journal of Physics: Conference Series **775** (2016) 012005 doi:10.1088/1742-6596/775/1/012005

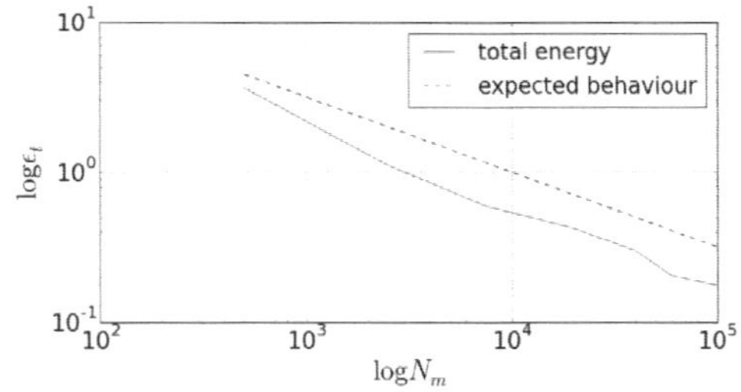

Figure 2: Convergence of the algorithm applied to the model (5)-(7). At $t = 1$, the error ϵ_t is plotted as a function of the number of markers N_m. The error decreases as $N_m^{-0.56}$ (the expected convergence rate is $O(1/\sqrt{N_m})$, cfr. black dashed curve). A similar result was found for what concerns the error on each mode, whenever the statistics is sufficient.

where $n_e(\rho)$ is the local equilibrium density, $F(\rho, \theta)$ is an envelope, $\rho = \sqrt{\psi}$, with ψ being the normalized poloidal flux of the tokamak equilibrium, and θ the geometrical poloidal angle. The functions ρ and θ are evaluated at the middle point $(x + x')/2$. At last the correlation is Gaussian in $x - x'$, with the matrix A accounting for the turbulence anisotropy along the equilibrium magnetic field direction. One has maximum flexibility in the choice of the envelope $F(\rho, \theta)$. Models used in this paper are sketched in Fig. 3.

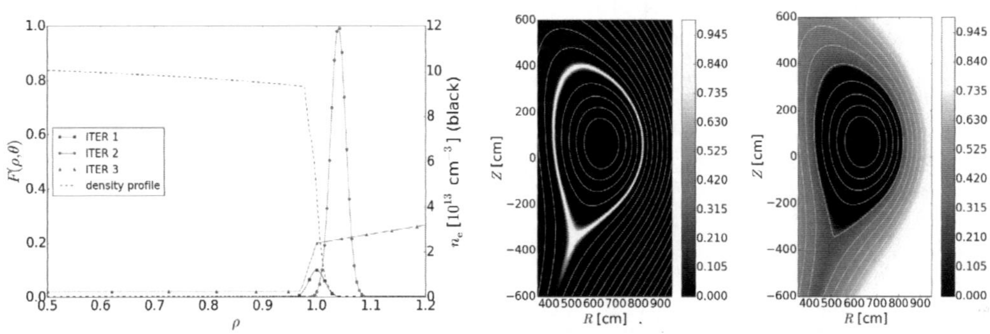

(a) Models for $F(\rho, \theta)$ (b) ITER 1 (c) ITER 3

Figure 3: Panel (a): models for the fluctuation envelope F as a function of ρ (we consider F uniform in θ), and density profile. Panels (b)-(c): contours of F in the poloidal plane for the considered ITER equilibrium. The inaccuracy in the reconstruction of the X-point region is an equilibrium post-processing artifact not affecting our simulations.

A possible choice for $F(\rho, \theta)$ is a Gaussian with center at $\rho = \rho_0$, amplitude $(\delta n_e/n_e)_0$, standard deviation $\delta\rho$, and uniform in θ. We observe different behaviours for different values of such parameters. In Table 1 the parameters used for the different simulations are listed, together with the amount of power absorbed at the resonance (considering 1.0 MW of injected power).

Scenario	$(\delta n_e/n_e)_0$	ρ_0	$\delta\rho$	Abs. power
ASDEX	0.1	1.0	0.02	1.0 MW
ITER 1	0.1	1.0	0.02	1.0 MW
ITER 2	1.0	1.04	0.02	0.64 MW

Table 1: Considered parameters for the case of a Gaussian turbulence envelope. The injected power is 1.0 MW in all cases.

In ASDEX the effect of fluctuations is actually negligible, as already pointed out in [7]. Fig. 4 (blue squares, red circles and green triangles) shows the power deposition profiles for ITER, together with the curve in absence of fluctuations for a comparison (black stars). All ITER scenarios refer to the upper launcher, low stearing mirror, with power deposition on the $q = 2$ resonant surface. The beam frequency is 170 GHz, and the correlation length of the fluctuations is 2cm in the direction perpendicular to the local magnetic field, while it is set to a large number in the parallel direction in order to take into account the flute-mode character of turbulence. For the scenario "ITER 1", cross-polarization scattering is negligible, as in the region where the fluctuations are located ($\rho = 1.0$) the density is high enough, and the dipersion relations of the modes are well separated [7]. Nonetheless, simple scattering to the same launched mode produces a significant broadening of the power deposition profile. In the scenario "ITER 2", the Gaussian layer of turbulence is moved far out in the scrape-off layer ($\rho = 1.04$), and its relative strength $(\delta n_e/n_e)_0$ is increased by a factor 10. This is an extremely severe fluctuation profile, which is articially manufactured in order to obtain a scenario in which a considerable fraction of the beam is scattered into the secondary mode (X-mode), and is afterwards reflected at the cutoff, located near the plasma separatrix. This produces a loss in terms of power deposited in the target region. On the other hand, the width of the power deposition region is not degraded much, in contrast to "ITER 1" where simple scattering is dominant.

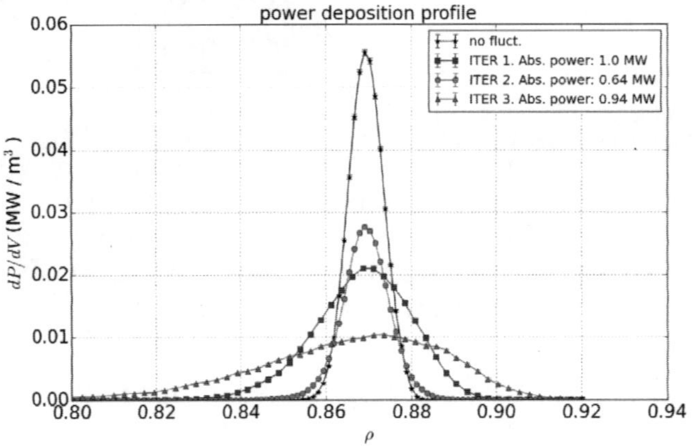

Figure 4: Power deposition profiles.

Together with the two standard Gaussian models, we have constructed a third scenario, ITER 3, in which $(\delta n_e/n_e)_0 = 0.02$ and the envelope is represented in Fig. 3a (green triangles curve). This should reflect better the expected shape of the correlation function at the edge of ITER plasmas. With such a model, about 6% of the injected power is scattered to the X-mode, which is reflected at the cut-off and does not reach the target resonance. Moreover, a significant broadening of the beam can also be observed (Fig. 4, ITER 3). A poloidal projection of the beam in this scenario is shown in Fig. 5.

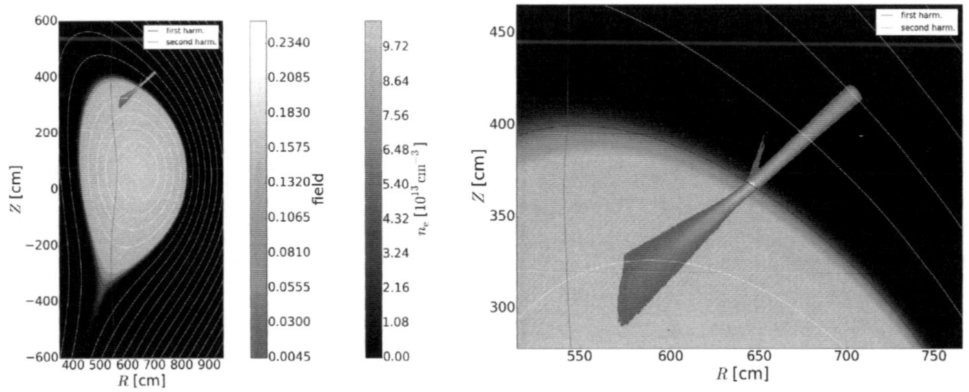

Figure 5: Scenario ITER 3 (poloidal section). The beam is injected from the upper launcher, lower stearing mirror, on the $q = 2$ resonance surface. About 6% of the power is scattered to the X-mode and reflected. A considerable broadening is observed for the O-mode near the deposition region. Again, the inaccuracy in the reconstruction of the X-point region is an equilibrium post-processing artifact which does not affect our simulations.

[1] Tsironis C *et al.* 2009 *Physics of Plasmas* **16** 112510
[2] Bertelli N *et al.* 2010 *Journal of Physics: Conference Series* **260**
[3] Decker J *et al.* 2012 *EPJ Web of Conferences* **32**
[4] Weber H 2013 *IPP Report* **5/134**
[5] McDonald S W 1991 *Phys. Rev. A* **43** 4484–4499
[6] McDonald S W 1988 *Physics Reports* **158** 337 – 416
[7] Weber H *et al.* 2015 *EPJ Web of Conferences* **87**
[8] Lapeyre B, Pardoux E and Sentis R 2003 *Introduction to Monte Carlo Methods for Transport and Diffusion Equations* (Oxford University Press)

This work has been carried out within the framework of the EUROfusion Consortium and has received funding from the European Union's Horizon 2020 research and innovation programme under grant agreement 633053. The views and opinions expressed herein do not necessarily reflect those of the European Commission.

Journal of Physics: Conference Series **775** (2016) 012006 doi:10.1088/1742-6596/775/1/012006

Padé approximation of the adiabatic electron contribution to the gyrokinetic quasi-neutrality equation in the ORB5 code

E. Lanti[1], J. Dominski[1], S. Brunner[1], B. F. McMillan[2], L. Villard[1]

[1] Ecole Polytechnique Fédérale de Lausanne (EPFL), Swiss Plasma Center (SPC), CH-1015 Lausanne, Switzerland

[2] Centre for Fusion, Space and Astrophysics, Department of Physics, Warwick University, Coventry, UK

E-mail: `emmanuel.lanti@epfl.ch`

Abstract. This work aims at completing the implementation of a solver for the quasi-neutrality equation using a Padé approximation in the global gyrokinetic code ORB5. Initially [Dominski, Ph.D. thesis, 2016], the Padé approximation was only implemented for the kinetic electron model. To enable runs with adiabatic or hybrid electron models while using a Padé approximation to the polarization response, the adiabatic response term of the quasi-neutrality equation must be consistently modified.

It is shown that the Padé solver is in good agreement with the arbitrary wavelength solver of ORB5 [Dominski, Ph.D. thesis, 2016]. To perform this verification, the linear dispersion relation of an ITG-TEM transition is computed for both solvers and the linear growth rates and frequencies are compared.

1. Introduction

It is well known that so-called anomalous transport due to micro-turbulence plays a dominant role in the particle and heat transport across the magnetic field lines of a magnetically confined plasma. This turbulence is mainly caused by small-scale instabilities such as the Ion Temperature Gradient (ITG) mode, the Trapped Electron Mode (TEM), and the Electron Temperature Gradient (ETG) mode that are driven by the plasma density and temperature gradients.

To describe efficiently these instabilities that typically evolve on slow time scales as compared to the ion cyclotron frequency, the gyrokinetic framework [1] is used. In this model, the fast gyro motion is averaged out, thus reducing the dimensionality of the phase space from 6D to 5D and eliminating fast time scales, resulting in a problem of decreased complexity. The distribution functions of each plasma species are evolved with the reduced set of gyrokinetic variables and the electrostatic field is solved consistently with the gyrokinetic Quasi-Neutrality Equation (QNE).

In the global gyrokinetic code ORB5, the field solver was originally using the Long Wavelength Approximation (LWA) for the polarization density [2, 3]. Recently, an Arbitrary Wavelength Solver (AWS) has been implemented based on previous works, see references [4, 5]. A preliminary version of a Padé solver has also been implemented which was only valid with the kinetic electron model [6, 7].

The Padé scheme involves approximating the polarization response with the inverse of a factor that includes the Laplacian; to make the resulting quasi-neutrality equation explicit, the whole equation is multiplied by this prefactor. The term in the quasi-neutrality equation involving adiabatic electron response must therefore also be consistently modified in this framework.

In this work, we have completed the former Padé solver by adding correction terms to the adiabatic electron response, allowing us to simulate perturbations down to small scales using the adiabatic electron model or a "hybrid" model in which trapped electrons are kinetic while passing electrons are assumed adiabatic. The complete Padé solver is of practical interest because it produces smaller matrices in terms of non-zero elements than the AWS which are, in turn, faster to assemble. As an example, for the runs performed during this work, the Padé solver required $1.5\,s$ to assemble the matrix whereas the AWS took $145\,s$. Furthermore, the Padé matrix needed ~ 9 times less memory to be stored than the AWS matrix. In the present version of the ORB5 code, the QNE is linearized, and the matrix needs to be computed only once. However, in view of possible extensions of the code, this performance gain can become crucial. For example, as we approach the tokamak edge region, the perturbation amplitude becomes comparable to the background equilibrium and the non-linear terms of the QNE need to be computed. This would require assembling the matrix at every time step of the simulation which is not practical with the AWS due to its long assembly time.

This paper is organized as follows. Section 2 describes the gyrokinetic QNE, as well as its long wavelength and Padé approximations. In Section 3, we present the discretization of the QNE and its implementation in the code. Finally, the code verification in the linear regime is presented in Section 4.

2. Computational Model

The linearized QNE for a two-species, electrostatic plasma can be written

$$\sum_{\sigma} q_\sigma \delta n_\sigma = q_i \delta n_i + q_e \delta n_e = q_i(\delta n_i^{gyr} + \delta n_i^{pol}) + q_e(\delta n_e^{gyr} + \delta n_e^{ad}) = 0 \tag{1}$$

where $\sigma = \{i, e\}$, q_σ are the electrical charges, $\delta n_i = \delta n_i^{gyr} + \delta n_i^{pol}$ and $\delta n_e = \delta n_e^{gyr} + \delta n_e^{ad}$. Here, $n_\sigma = n_{0\sigma} + \delta n_\sigma$ with $n_{0\sigma}$ being the background density, δn_σ the perturbed density. The background densities are assumed to satisfy the QNE, i.e. $q_i n_{i0} + q_e n_{e0} = 0$.

In Eq. (1), δn_e^{ad} is the adiabatic contribution to the perturbed electron density given by

$$\delta n_e^{ad}(\vec{x}, t) = -\alpha_{ad} \frac{q_e n_{e,0}}{T_{e,0}} \left[\delta\phi(\vec{x}, t) - \delta\bar{\phi}(s, t)\right], \tag{2}$$

where $\vec{x} = (s, \theta^\star, \varphi)$ with s, $\theta^\star$, and φ being respectively the radial, poloidal and toroidal coordinates, $T_{e,0}$ is the equilibrium electron temperature, $\delta\bar{\phi}$ is the flux-surface averaged perturbed potential, and α_{ad} is the fraction of adiabatic electrons ($\alpha_{ad} = 1$ for the adiabatic electron model, $\alpha_{ad} = 0$ for the fully kinetic electron model, and α_{ad} is equal to the fraction of passing electrons for the hybrid model). Both δn_i^{gyr} and δn_e^{gyr} are the ion and electron gyro densities:

$$\delta n_i^{gyr}(\vec{x}, t) = \int J \mathrm{d}\vec{Z} \delta(\vec{R} + \vec{\rho_i} - \vec{x}) \delta f_i, \qquad \delta n_e^{gyr}(\vec{x}, t) = \int J \mathrm{d}\vec{Z} \delta(\vec{R} - \vec{x}) \delta f_e, \tag{3}$$

where δf_σ is the perturbed distribution function, J the Jacobian relative to the transformation from particle to guiding-center variables, $\vec{R}$ the guiding-center position, $\vec{\rho_i}$ the ion Larmor radius, and finally $\mathrm{d}\vec{Z}$ the infinitesimal phase-space volume. Note that, in the last integral of Eq. (3),

Journal of Physics: Conference Series **775** (2016) 012006 doi:10.1088/1742-6596/775/1/012006

only the phase-space region of non-adiabatic electrons is considered. Finally, the gyrokinetic ion polarization density is given by

$$n_i^{pol}(\vec{x},t) = \int J\mathrm{d}\vec{Z}\,\delta(\vec{R}+\vec{\rho}_i-\vec{x})\,\frac{q_i}{B_0}\delta\tilde{\phi}\frac{\partial f_{i,0}}{\partial\mu}, \tag{4}$$

where B_0 is the equilibrium magnetic field amplitude, and $\delta\tilde{\phi} = \delta\phi - \langle\delta\phi\rangle_\alpha$ with $\langle\delta\phi\rangle_\alpha$ being the gyro-averaged electrostatic potential.

In ORB5, the quasi-neutrality equation (1) with ion polarization density given by (4) is solved in its integral form by the AWS recently implemented, see references [6, 7]. This upgrade was added in order to overcome the original LWA made in the code, in which the ion polarization density contribution is given by

$$\delta n_i^{pol} \simeq \frac{q_i}{m_i}\nabla_\perp\cdot\frac{n_{i,0}}{\Omega_{i,0}^2}\nabla_\perp\delta\phi, \tag{5}$$

where $\Omega_{i,0} = q_iB_0/m_i$ and m_i is the ion mass. In the Padé approximation [4, 8], one considers the improved approximation

$$\delta n_i^{pol} \simeq \left[1-\nabla_\perp\cdot\rho_i^2\nabla_\perp\right]^{-1}\left[\frac{q_i}{m_i}\nabla_\perp\cdot\frac{n_{i,0}}{\Omega_{i,0}^2}\nabla_\perp\delta\phi\right]. \tag{6}$$

At this point, all the QNE is multiplied by $\left[1-\nabla_\perp\cdot\rho_i^2\nabla_\perp\right]$ which results in

$$\frac{q_i}{m_i}\nabla_\perp\cdot\frac{n_{i,0}}{\Omega_{i,0}^2}\nabla_\perp\delta\phi + \delta n_e^{ad}(\vec{x},t) - \nabla_\perp\cdot\rho_i^2\nabla_\perp\delta n_e^{ad}(\vec{x},t) = \tag{7}$$

$$\delta n_i^{gyr}(\vec{x},t) - \nabla_\perp\cdot\rho_i^2\nabla_\perp\delta n_i^{gyr}(\vec{x},t) + \delta n_e^{gyr}(\vec{x},t) - \nabla_\perp\cdot\rho_i^2\nabla_\perp\delta n_e^{gyr}(\vec{x},t). \tag{8}$$

In Eq. (7), the first term is the LWA of the ion polarization density, the second term is the adiabatic electron density contribution and the third term is the Padé correction to the adiabatic electron density contribution. In Eq. (8), the first and third terms are respectively the ion and electron gyro densities, and the second and third terms are the Padé corrections to these gyro densities.

The first version of the Padé solver, already implemented in ORB5, was actually only valid for the kinetic electron model [6]. In the present work, we are thus interested in implementing the Padé correction terms related to the adiabatic electron contribution to the QNE. These corrections terms are necessary for both the adiabatic and the hybrid electron models.

3. Discretization of the QNE equation

In ORB5, the QNE is discretized using a finite-element representation of the electrostatic field

$$\delta\phi(s,\theta^\star,\varphi) = \sum_{\nu'}\delta\phi_{\nu'}\Lambda_{\nu'}(s,\theta^\star,\varphi), \tag{9}$$

where $\Lambda_{\nu'}(s,\theta^\star,\varphi) = \Lambda_{i'}(s)\Lambda_{j'}(\theta^\star)\Lambda_{l'}(\varphi)$ are a tensor product of one-dimensional B-splines basis functions [2, 9]. With this representation the QNE is discretized into a linear system of equations following the Ritz-Galerkin method:

$$\sum_{\nu'}L_{\nu\nu'}\delta\phi_{\nu'} = S_\nu. \tag{10}$$

Journal of Physics: Conference Series **775** (2016) 012006 doi:10.1088/1742-6596/775/1/012006

In the Padé approximation, similarly to equations 7 and 8, the LHS matrix, $L_{\nu\nu'} = L^{LWA}_{\nu\nu'} + L^{ad}_{\nu\nu'} + L^{ad,corr}_{\nu\nu'}$, is composed of the LWA of the polarization density $L^{LWA}_{\nu\nu'}$, the adiabatic electron response $L^{ad}_{\nu\nu'}$, and the Padé correction to the adiabatic response $L^{ad,corr}_{\nu\nu'}$. The source term in the RHS, $S_\nu = S^{i,gyr}_\nu + S^{e,gyr}_\nu + S^{i,Corr}_\nu + S^{e,Corr}_\nu$, is composed of the ion and electron gyro densities contributions, $S^{i,gyr}_\nu$ and $S^{e,gyr}_\nu$, and the Padé corrections to the gyro densities, $S^{i,Corr}_\nu$ and $S^{e,Corr}_\nu$. These two corrections terms have already been implemented for compatibility with the kinetic electron model. To handle both the adiabatic and hybrid electron models, the additional correction term to the adiabatic electron response has been added in the code. It is of the form $L^{ad,corr}_{\nu\nu'}\,\delta\phi_{\nu'}$ with

$$\begin{aligned} L^{ad,corr}_{\nu\nu'} &= -\int \mathrm{d}\vec{x}\Lambda_\nu(\vec{x})\nabla_\perp\cdot\rho_i^2(\vec{x})\nabla_\perp\left\{\frac{q_e^2 n_{e,0}(s)}{T_{e,0}(s)}\left[\Lambda_{\nu'}(\vec{x},t)-\bar{\Lambda}_{\nu'}(s)\right]\right\} \\ &= \int \mathrm{d}\vec{x}\rho_i^2(\vec{x})\nabla_\perp\Lambda_\nu(\vec{x})\cdot\left[\frac{q_e^2 n_{e,0}(s)}{T_{e,0}(s)}\right]\nabla_\perp\left[\Lambda_{\nu'}(\vec{x},t)-\bar{\Lambda}_{\nu'}(s)\right] \\ &\quad \underbrace{+\int \mathrm{d}\vec{x}\rho_i^2(\vec{x})\nabla_\perp\Lambda_\nu(\vec{x})\cdot\nabla_\perp\left[\frac{q_e^2 n_{e,0}(s)}{T_{e,0}(s)}\right]\left[\Lambda_{\nu'}(\vec{x},t)-\bar{\Lambda}_{\nu'}(s)\right]}_{\text{neglected}}. \end{aligned}$$

The last term of this equation, depending on the equilibrium profile gradient, is not considered in this work. It can be shown that it is $\varepsilon = \rho_i/L_{T,n}$ smaller than the first term, and can therefore be neglected.

Because of the toroidal and poloidal periodicity, the discretized QNE can be Fourier transformed using a DFT. The advantage of this procedure is multiple. First, as the system is toroidally axisymmetric, the toroidal mode numbers n are decoupled [2]:

$$\sum_{\nu'} L^{(n)}_{\nu\nu'}\delta\phi_{\nu'} = S^{(n)}_\nu. \tag{11}$$

Furthermore, based on a work presented in Ref. [10], only the relevant poloidal modes are assembled. Indeed, since the turbulence of interest is principally aligned with the magnetic field, then for each toroidal mode number n only a few poloidal mode numbers $m = nq \pm \Delta m$ need to be retained (for most simulations $\Delta m = 5$ is enough to converge the results), where q is the safety factor. This can be used during the assembly in order the reduce the size of the matrix and thus, speeding up the computations.

4. Code Verification in the linear regime

The new version of the Padé solver is verified in the linear regime against the AWS which was benchmarked [6] against the GENE code [11]. To this end, a modified CYCLONE test case [12] is used to simulate an ITG-TEM transition. The physical parameters are: $\rho^\star = \rho_i/a = 1/90$, where $a = 0.61\,[m]$ is the minor radius, $R_0 = 1.70\,[m]$ is the major radius, $B_0 = 1.91\,[T]$ is the magnetic field amplitude, $q_0 = 0.85$ and $q_{edge} = 3.04$ are the values of the safety factor at the axis and at the edge with $q(s) = q_0 + (q_{edge} - q_0)s^2$, $m_i/m_e = 20$, $q_i/q_e = -1$. The temperature and density profiles are defined by

$$\frac{T_\sigma(\tilde{\rho})}{T_\sigma(\tilde{\rho}_{0\sigma})} = \exp\left[-\frac{a\Delta_{T_\sigma}}{L_{T_\sigma}}\tanh\left(\frac{\tilde{\rho}-\tilde{\rho}_{0\sigma}}{\Delta_{T_\sigma}}\right)\right], \quad \frac{n(\tilde{\rho})}{n(\tilde{\rho}_0)} = \exp\left[-\frac{a\Delta_n}{L_n}\tanh\left(\frac{\tilde{\rho}-\tilde{\rho}_0}{\Delta_n}\right)\right] \tag{12}$$

where $\tilde{\rho} = \rho/a$ is a normalized radial coordinate, $\tilde{\rho}_{0\sigma} = 0.5$, $\Delta_{T_\sigma} = \Delta_n = 0.30$, $T_i = T_e$, $L_n/L_{T_i} = 3.14$ is the density to temperature characteristic length ratio, and $R_0/L_{T_i} = 6.92$.

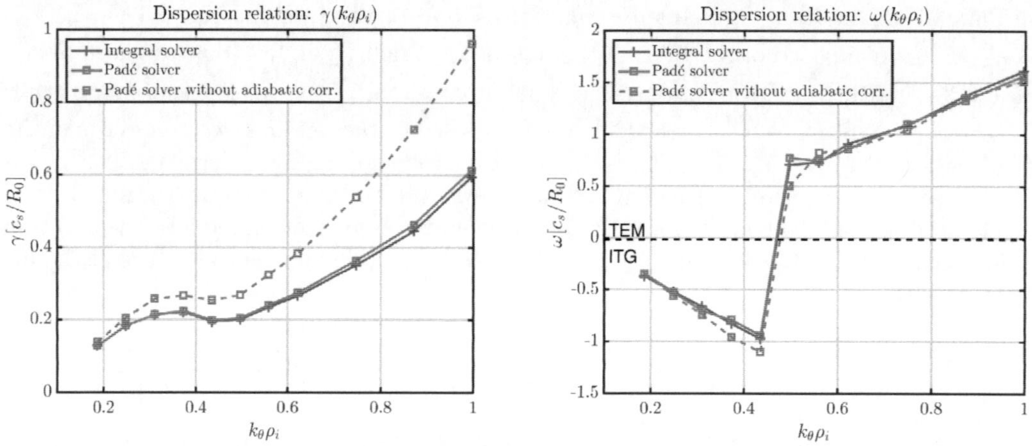

Figure 1. Dispersion relation of both ITG and TEM modes obtained with ORB5's Padé and AWS solvers for a CYCLONE-like test case. The left figure shows the linear growth rate of the dominant mode as a function of $k_\theta\rho_i$. The right figure shows the linear frequencies of these modes.

The numerical parameters are the following: $N_M = 2 \cdot 10^6$ is the number of numerical markers, $\Delta t = 16\Omega_i^{-1}$ is the time step, $n_s = 120$, $n_\theta^\star = 512$ and $n_\varphi = 128$ are the number of grid intervals in the s, $\theta^\star$ and φ directions, the simulation domain is annular from $s = 0.3$ to $s = 1$ and the hybrid electron model is used.

The simulation results are plotted in Figure 1, where the frequency and growth rate are shown as a function of $k_\theta\rho_i$, where k_θ is the poloidal mode wave number.

The left plot shows the linear growth rates of the most unstable modes. The Padé solver (solid green line) is in very good agreement with the AWS (solid blue line). For values $k_\theta\rho_i > 0.6$, the Padé solver slightly overestimates the growth rate but the difference is less than 3.5%. To emphasize the importance of the Padé correction to the adiabatic electron contribution, a scan is made neglecting $L_{\nu\nu'}^{ad,corr}$, i.e. only the drift-kinetic electrons have the Padé correction. It is seen that the growth rate is always overestimated, especially for $k_\theta\rho_i > 0.4$. The right plot of Figure 1 shows the linear frequencies of the modes for a similar scan in $k_\theta\rho_i$. The Padé solver is also in agreement with the AWS. More surprisingly, the Padé correction to the adiabatic electron contribution seems to be negligible for the estimation of the frequency.

In Figure 2, we represent a poloidal cut for the three solvers: Long Wavelength Solver (LWS), AWS and Padé. For all cases, the cut shows the electrostatic potential and is taken at $t = 8000\,\Omega_{ci}^{-1}$ and for $k_\theta\rho_i = 0.62$.

The perturbation computed by the Padé solver (right) has a slightly higher amplitude due to the growth rate that is slightly overestimated by Padé as compared to the AWS (center) ($\gamma_{Pade} = 0.27\,c_s/R_0$ and $\gamma_{AWS} = 0.26\,c_s/R_0$). On the top-left of each figure is shown a zoom of the poloidal mode structure. Both Padé solver and AWS resolve correctly the fine structures (radial ripples in the zoomed pictures). On the other hand, the LWS (left) fails at solving the fine structures and the amplitude of the perturbation is around ten times smaller than the AWS because the LWS largely underestimates the growth rate: we are beyond the validity limit of the long wavelength approximation.

Thanks to this new adiabatic electron correction term, the Padé solver allows one to solve for short-scales instabilities that arise around mode rational flux surfaces when using both the adiabatic and hybrid electron models.

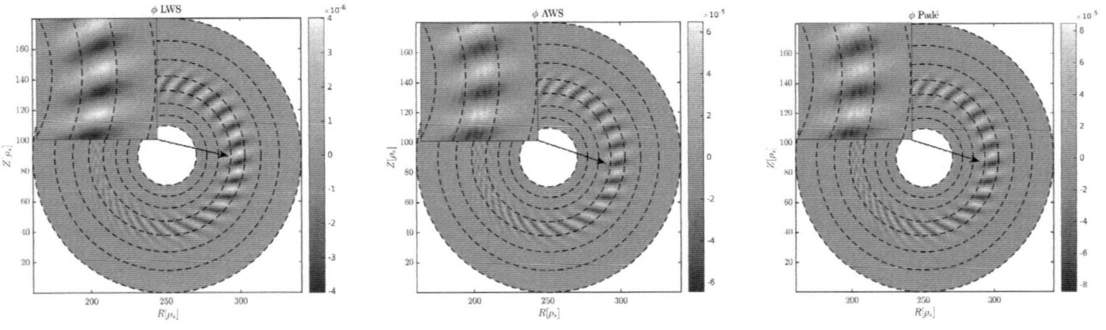

Figure 2. Poloidal cuts for $t = 8000\,\Omega_{ci}^{-1}$ and $k_\theta\rho_i = 0.62$ for all three cases. The LWS is on the left, the AWS in the center and the Padé version of the solver on the right.

5. Conclusions

The new adiabatic electron correction term for the Padé version of the solver has been successfully implemented in ORB5 for $n \neq 0$ mode numbers and allows one to accurately simulate the ITG-TEM transition. Developments related to the $n = 0$ modes are under progress, which are necessary for the simulation of turbulence with the adiabatic electron approximation when using the Padé version of the solver. Other tests (e.g. Rosenbluth-Hinton [13]) are being carried out to finalize the verification of the Padé solver in the linear regime.

Linear dispersion relation (growth rate and frequency) shows that Padé solver has a good agreement with the AWS. For $k_\theta\rho_i > 0.6$, Padé solver slightly overestimates the growth rate as compared to AWS. Nevertheless, the difference is always smaller than 3.5%. The contribution of the Padé correction to the adiabatic electron response is shown to be non-negligible for the growth rate estimation. The mode frequencies obtained with both Padé solver and AWS are in good agreement even without the Padé correction to the adiabatic response.

Finally, we have shown that the Padé solver, as for the AWS, is capable of solving the fine structures that arise around mode rational surfaces.

6. Acknowledgments

This work was partly supported by the PASC project and by the Swiss National Science Foundation. This work has been carried out within the framework of the EUROfusion Consortium and has received funding from the Euratom research and training programme 2014-2018 under grant agreement No 633053. The views and opinions expressed herein do not necessarily reflect those of the European Commission.

References

[1] Brizard A J and Hahm T S 2007 *Reviews of Modern Physics* **79** 421–468 ISSN 0034-6861 URL `http://journals.aps.org/rmp/abstract/10.1103/RevModPhys.79.421`

[2] Tran T M, Appert K, Fivaz M, Jost G, Vaclavik J and Villard L 1999 *Theory of fusion plasmas* (Societa Italiana di Fisica) p 45

[3] Jolliet S, Bottino A, Angelino P, Hatzky R, Tran T, McMillan B F, Sauter O, Appert K, Idomura Y and Villard L 2007 *Computer Physics Communications* **177** 409–425 ISSN 00104655 URL `http://www.sciencedirect.com/science/article/pii/S0010465507002251`

[4] Lin Z and Lee W W 1995 *Physical Review E* **52** 5646–5652 ISSN 1063-651X URL `http://link.aps.org/doi/10.1103/PhysRevE.52.5646`

[5] Mishchenko A, Konies A and Hatzky R 2005 *Physics of Plasmas* **12** 062305 ISSN 1070664X URL `http://scitation.aip.org/content/aip/journal/pop/12/6/10.1063/1.1925587`

[6] Dominski J 2016 *To be submitted to Physics of Plasmas*

[7] Dominski J 2016 *Development and application of Eulerian and Particle-In-Cell gyrokinetic codes for studying the effect of non-adiabatic passing electron dynamics on microturbulence* Ph.D. thesis École Polytechnique Fédérale de Lausanne

[8] Bottino A 2004 *Modelling of global electrostatic microinstabilities in tokamaks* Ph.D. thesis École Polytechnique Fédérale de Lausanne

[9] Fivaz M, Brunner S, de Ridder G, Sauter O, Tran T, Vaclavik J, Villard L and Appert K 1998 *Computer Physics Communications* **111** 27–47 ISSN 00104655 URL `http://linkinghub.elsevier.com/retrieve/pii/S001046559800023X`

[10] McMillan B, Jolliet S, Bottino A, Angelino P, Tran T and Villard L 2010 *Computer Physics Communications* **181** 715–719 ISSN 00104655 URL `http://www.sciencedirect.com/science/article/pii/S0010465509003920`

[11] Jenko F, Dorland W, Kotschenreuther M and Rogers B N 2000 *Physics of Plasmas* **7** 1904 ISSN 1070664X URL `http://scitation.aip.org/content/aip/journal/pop/7/5/10.1063/1.874014`

[12] Dimits A M, Bateman G, Beer M a, Cohen B I, Dorland W, Hammett G W, Kim C, Kinsey J E, Kotschenreuther M, Kritz a H, Lao L L, Mandrekas J, Nevins W M, Parker S E, Redd a J, Shumaker D E, Sydora R and Weiland J 2000 *Physics of Plasmas* **7** 969 ISSN 1070664X URL `http://scitation.aip.org/content/aip/journal/pop/7/3/10.1063/1.873896`

[13] Rosenbluth M N and Hinton F L 1998 *Physical Review Letters* **80** 724–727 ISSN 0031-9007 URL `http://link.aps.org/doi/10.1103/PhysRevLett.80.724`

Characterization with microturbulence simulations of the zero particle flux condition in case of a TCV discharge showing toroidal rotation reversal

A. Mariani, G. Merlo, S. Brunner, A. Merle, O. Sauter
Swiss Plasma Center, EPFL, 1015 Lausanne, Switzerland
E-mail: alberto.mariani@epfl.ch

T. Görler
Max-Planck-Institut für Plasmaphysik, Boltzmannstraße 2, D-85748 Garching, Germany

F. Jenko, D. Told
UCLA Physics & Astronomy, 475 Portola Plaza, Los Angeles, USA

Abstract. In view of the stabilization effect of sheared plasma rotation on microturbulence, it is important to study the intrinsic rotation that develops in tokamaks that present negligible external toroidal torque, like ITER. Remarkable observations have been made on TCV, analysing discharges without NBI injection, as reported in [A. Bortolon et al. 2006 Phys. Rev. Lett. 97] and exhibiting a rotation inversion occurring in conjunction with a relatively small change in the plasma density. We focus in particular on a limited L-mode TCV shot published in [B. P. Duval et al. 2008 Phys. Plasmas 15], that shows a rotation reversal during a density ramp up. In view of performing a momentum transport analysis on this TCV shot, some constraints have to be considered to reduce the uncertainty on the experimental parameters. One useful constraint is the zero particle flux condition, resulting from the absence of direct particle fuelling to the plasma core. In this work, a preliminary study of the reconstruction of the zero particle flux hyper-surface in the physical parameters space is presented, taking into account the effect of the main impurity (carbon) and beginning to explore the effect of collisions, in order to find a subset of this hyper-surface within the experimental error bars. The analysis is done performing gyrokinetic simulations with the local (flux-tube) version of the Eulerian code GENE [Jenko et al 2000 Phys. Plasmas 7 1904], computing the fluxes with a Quasi-Linear model, according to [E. Fable et al. 2010 PPCF 52], and validating the QL results with Non-Linear simulations in a subset of cases.

1. Introduction

It is widely accepted that sheared flows can play a role in stablizing microturbulence, potentially decreasing the transport to neoclassical levels [1] and contributing to the formation of transport barriers which in particular provide access to the H-mode and advanced operating regimes, showing a better plasma confinement with respect to the basic L-mode discharges [2]. The simplest way to produce a net toroidal plasma rotation in the tokamak plasma core is to submit the system to an external torque by injecting NBI beams with a non-vanishing toroidal momentum. This momentum can be exchanged with the plasma bulk species by mainly three processes, that is collisions, formation of a radial electric field by the NBI neutrals ionisation and direct inclusion of the ionised beam neutrals in the bulk species due to magnetic confinement [3]. Even without the presence of an external torque, tokamak plasma can autonomously develop a net toroidal rotation [4]. In this case the rotation is called 'intrinsic'. The observed intrinsic toroidal rotation in plasmas with no externally applied torque could determine a possibility of sustaining the rotation profile in next step devices, where the toroidal momentum driven by external systems may not dominate that generated by the plasma itself. In particular, the study of intrinsic rotation is pivotal to understand the behaviour of plasma transport and confinement in the planned ITER scenarios, where it is expected a negligible external torque, due to the limited penetration depth of NBI [5].

The TCV tokamak at SPC Lausanne [6] is a machine particularly suited for investigating intrinsic rotation. It features a low power diagnostic-NBI system, of negligible torque, which is used in conjunction with Charge Exchange Recombination Spectroscopy (CXRS) measurement to reconstruct ion density and velocity profiles. In particular, remarkable observations have been made on TCV, as reported in [7], exhibiting an intrinsic rotation inversion in the plasma core for L-mode discharges, occurring in conjunction with a relatively small change in the plasma density. A possible explanation for this behaviour has been suggested in form of a transition in the microturbulence with increasing density from a dominantly Trapped Electron Mode (TEM) to a dominantly Ion Temperature Gradient (ITG) regime [8]. Indeed, the TEMs can be significantly damped by collisions (increasing with density) or other processes correlated to density increase, leading to this transition. The change of sign of the residual part of the toroidal momentum radial flux (the part that is proportional neither to the toroidal velocity nor to its radial derivative) as a consequence of a mode population change from TEM to ITG is invoked to possibly explain the reversal [9].

In this work, we focus on the analysis of the limited L-mode TCV shot # 28355, published in [10], that shows a rotation reversal during a density ramp up. Gyrokinetic codes, which enable us to simulate the turbulent transport in magnetic confinement plasmas, should provide a powerful tool to study the micro turbulent regimes present in the plasma core close to the toroidal rotation inversion, enabling to test the hypothesis of the TEM-ITG transition as a possible explanation. The analysis of the momentum transport in the considered TCV shot is complicated by the fact that the magnetic equilibrium is with a good approximation up-down symmetric. Indeed, it is known that each contribution to the residual part of the toroidal momentum flux is due to a 'symmetry breaking' mechanism [11], and recently in [12] it has been shown that only the dominant contribution in a $\rho^\star$ expansion ($\rho^\star = \rho_i/a$, where ρ_i is the ion larmor radius and a the tokamak minor radius), due to the up-down asymmetry of the magnetic surfaces, can be obtained using standard flux-tube gyrokinetic codes (to obtain other contributions to the residual toroidal momentum flux, the gyrokinetic equations have to be written at an higher order in $\rho^\star$ than the one implemented in standard local codes). Nevertheless, a preliminary study can be performed, by means of linear and Non-Linear (NL) gyrokinetic simulations, to examine the turbulence regime close to the rotation transition.

In view of a momentum transport analysis planned to be carried on this shot, since the main physical parameters, that is the density and temperature gradients, the electron to ion temperature ratio, the main impurity concentration and gradient, have big experimental uncertainty, it is required to find constraints to reduce the freedom in their choice. Since in the considered discharge there is no external particle injection, it is expected that at every magnetic surface the average particle flux is vanishing for every plasma species. Identifying the physical parameters satisfying this constraint, resulting in a hyper-surface in the parameter space, and in particular the evaluation of the electron density gradient corresponding to the vanishing particle fluxes, is in itself a topic of research [13].

This work is accomplished performing linear flux-tube gyrokinetic transport simulations with the local version of the GENE code [15], computing the fluxes with a Quasi-Linear (QL) model and validating the results in a limited number of cases with NL simulations. We consider for simplicity mainly the collisionless regime, aiming to perform the same analysis adding collisions, isolating their effect on the results. The effect of the main impurity (carbon) on the zero particle flux condition is investigated, with the goal of finding a result consistent, within error bars, with experimental parameters. As will be shown, the obtained result turns out to be consistent with a turbulence regime where TEM and ITG modes are balanced throughout the k_y spectrum. This is consistent with the standard explanation of the mechanism behind the rotation reversal, namely a transition from TEM to ITG turbulence. Then we present a preliminary result on the effect of collisions, disregarding the effect of carbon, showing that in the collisional regime the obtained density peaking factor is not consistent with the experimental one anymore. Nevertheless, relevant parameters such as the electron temperature gradient and the electron to ion temperature ratio have still to be scanned in the collisional case, and their effect could maybe compensate the effect of collisions in the density peaking factor determination.

2. Results & Discussion

2.1. Input Parameters

The magnetic equilibrium and plasma parameters considered in this work are the ones relative to TCV shot # 28355, at $t = 0.96$ s, just before the rotation reversal. This shot, a limited ohmic L-mode one, presents a deuterium plasma, with carbon as main impurity. Temperature and density profiles of all species, as well as the magnetic geometry, are taken from experimental measurement. The magnetic equilibrium is reconstructed using the ideal MHD solver CHEASE [16]. In Fig. 1 the electron density profile $n_e(\rho_{\rm tor,n})$ and the carbon toroidal velocity profile $v_{tor,c}(\rho_{\rm tor,n})$ are plotted versus the normalised toroidal radius $\rho_{\rm tor,n} \equiv \rho_{\rm tor}/\rho_{\rm tor,edge}$, where $\rho_{\rm tor} = \sqrt{\Phi/\pi B_0}$, with Φ the toroidal flux and $B_0 \simeq 1.44T$ the vacuum field at torus major radius, before and after the rotation inversion. The sign of the toroidal velocity is relative to the plasma current. Therefore the rotation reversal, as can be seen, is from counter-current to co-current.

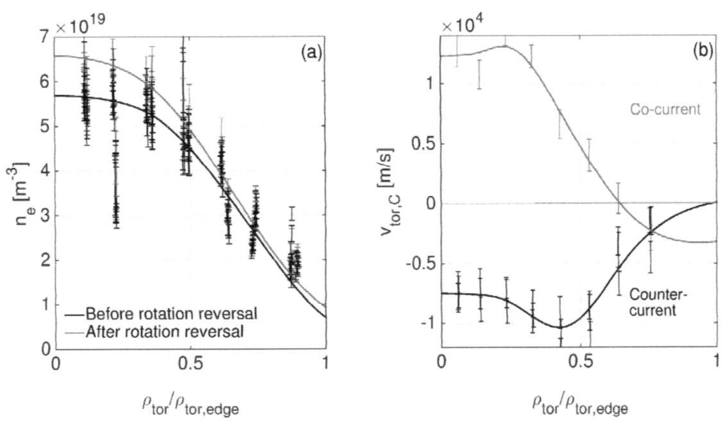

Figure 1. (a) Electron density radial profile before (black, average over $t \in [0.85, 1.06]$ s) and after (red, average over $t \in [1.12, 1.15]$ s) the rotation inversion for TCV shot # 28355; (b) Carbon toroidal velocity radial profile before and after the reversal.

The analysis has been performed at the radial position $\rho_{\rm tor,n} = 0.6$, which lies inside the 'stiff' region for the density and temperature profiles $\rho_{\rm inv} < \rho_{\rm vol,n} \sim \rho_{\rm tor,n} < 0.8$, where $\rho_{\rm inv}$ is the sawteeth 'inversion radius' [17]. In this radial interval it is particularly important to reduce the experimental uncertainty on the density and temperature gradients, because the fluxes are very sensitive to their variation.

The experimental density and temperature radial profiles are shown in Fig. 2 (a) and (b) for the main species for time $t = 0.957\ s$, before reversal. The electron profiles have been obtained with the Thomson scattering diagnostic, while the carbon profiles have been measured with CXRS. The ion density profile is computed enforcing neutrality $\sum_{sp} q_{sp} n_{sp} = n_i + 6n_c - n_e = 0$, while the ions and the carbon are assumed to be in thermal equilibrium $T_i = T_c$. In Fig. 2 (c) and (d) the safety factor, shear and $Z_{\rm eff} = \sum_j Z_j^2 n_j / \sum_j Z_j n_j = (n_i + 36n_c)/n_e$ (j indicates the ion species) profiles are presented.

The main experimental parameters used in the gyrokinetic simulations at $\rho_{\rm tor,n} = 0.6$ are summarized in Table 1. Here, $R/L_f \equiv -(R/\rho_{\rm tor,edge}) d\ln f / d\rho_{\rm tor,n}$ represents the normalised 'radial' logarithmic gradient of the f profile evaluated at $\rho_{\rm tor,n} = 0.6$, where $R = 0.88m$ is the major radius of the torus, and $\bar{\nu} = \nu_{ei} R/c_s$ is the normalised electron-ion collision frequency, with

$$\nu_{e,i} = \sqrt{2}\pi \frac{e^4}{(4\pi\epsilon_0)^2} \frac{n_i \ln\Lambda}{m_e^{1/2} T_e^{3/2}} \tag{1}$$

the electron-ion collision frequency in Hz (being all the quantities appearing in its definition in S.I. units), where $\ln\Lambda$ is the Coulomb logarithm ($\ln\Lambda \sim 10 - 15$), and $c_s \equiv \sqrt{T_e/m_i}$ the Deuterium sound speed. Moreover, q is the safety factor, $\hat{s}$ is the magnetic shear, and $\beta \simeq 8\pi n_e T_e / B_0^2$ is the plasma beta. The experimental error bars are estimated to be of the order of $\pm 20\%$ on the profiles and roughly $\pm 40\%$ on the normalised logarithmic gradients.

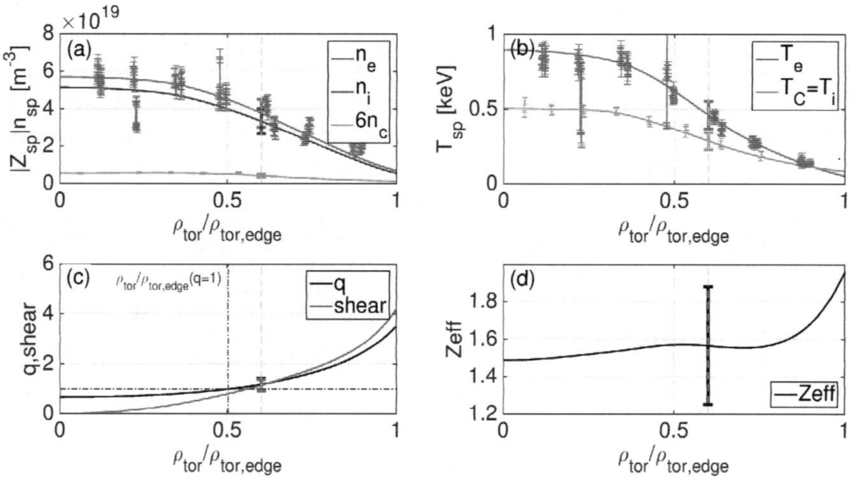

Figure 2. (a) Density radial profiles for electrons, deuterons and carbon, multiplied by their respective ionisation degree $|Z_{sp}|$, and satisfying neutrality $n_e = n_i + 6n_c$; (b) Temperature profiles for electrons and carbon. Ions (deuterons) are assumed in thermal equilibrium with carbon; (c) Safety factor and magnetic shear profiles; (d) $Z_{\rm eff}$ profile.

In the 5-dimensional reduced gyrokinetic phase space, GENE adopts a field-aligned coordinate system (x, y, z) in the configuration space, while $(v_\parallel, \mu)$ are used as velocity variables. Here (x, y, z) represent the radial, the binormal and the parallel ("parallel" as it follows the position

Table 1. Main experimental parameters at $\rho_{\text{tor,n}} = 0.6$

R/L_{ne}	R/L_{ni}	R/L_{nc}	R/L_{Te}	R/L_{Ti}	T_i/T_e	$\bar{\nu}$	Z_{eff}	q	$\hat{s}$	β
6.15	6.05	6.9	9.6	8	0.62	1.26	1.57	1.19	1.15	0.34×10^{-2}

along a given magnetic field line) positions respectively, $\mu = mv_\perp^2/2B$ is the magnetic moment and $v_\parallel$ is the component of the velocity in the magnetic field direction. In the flux-tube version of the code, Fourier representation is used for the x and y directions. A typical grid size for a linear simulation of fixed mode number k_y wrt. y direction ($k_y = nq/\rho_{\text{tor,n}}$, where n is the toroidal mode number) is $n_{kx} \times n_z \times n_{v\parallel} \times n_\mu = 48 \times 32 \times 48 \times 16$. Convergence tests have been performed to check the reliability of the results. In all the simulations, the electrons have been treated as self-consistent gyrokinetic species.

2.2. Fluxes and Quasi-Linear model

Aiming to find the zero particle flux hyper-surface in the multi-dimensional space of parameters characterising the physical system, the radial particle flux is efficiently estimated with a Quasi-Linear (QL) model, considering only the electrostatic contribution (since β is small, as can be seen in Table 1), according to [13]:

$$\Gamma^{QL} = A_0 \sum_{k_y} \left(\frac{\gamma_{k_y}}{\langle k_\perp^2 \rangle (k_y)} \right)^\xi \frac{\Gamma^L(k_y)}{|\hat{\phi}(k_y, k_x = 0, z = 0)|^2} \ . \tag{2}$$

Here, the average squared perpendicular wave number is defined by

$$\langle k_\perp^2 \rangle (k_y) = \sum_{k_x} \int (g^{xx} k_x^2 + 2g^{xy} k_x k_y + g^{yy} k_y^2) |\hat{\phi}(k_x, k_y, z)|^2 dz \Big/ \sum_{k_x} \int |\hat{\phi}(k_x, k_y, z)|^2 dz \ , \tag{3}$$

with γ_{k_y} the growth rate of the most unstable mode, $\hat{\phi}$ the x, y Fourier transform of the electrostatic potential, and $\{g^{ij}\ ,\ j = x, y, z\}$ the double-contravariant components of the metric tensor. The A_0 constant is a scaling factor associated to the absolute flux magnitude, which does not need to be determined as we are interested in the zero particle condition only, which can be obtained considering the ratio Γ/Q, where Q is the electrostatic part of the radial heat flux, following [14]. The exponent ξ is determined with the help of few nonlinear simulations (see section 2.6). In all cases, unless differently stated we have used $\xi \equiv 2$, in agreement with [13], and this assumption has been justified comparing QL and NL simulations. Finally, $\Gamma^L(k_y)$ is the k_y spectral contribution of the k_y mode to the electrostatic part of the linear radial particle flux, satisfying

$$\Gamma^L = \sum_{k_y} \Gamma^L(k_y) \tag{4}$$

$$\Gamma^L = \langle n v_x \rangle = -\langle n \partial_y \phi \rangle \tag{5}$$

the electrostatic part of the linear radial particle flux, where n is the considered species density, v_x is the covariant component of the velocity, ϕ is the electrostatic potential and $\langle \ldots \rangle$ indicates a spatial averaging over the simulation domain. Following [18], the x, y part of the spatial average, since it involves a product of two quantities, is performed spectrally by applying Parseval Theorem, leading to the final expression of the k_y spectrum:

$$\Gamma^L(k_y) = \int dz \sum_{k_x} 2\Re\left\{[\hat{n}(k_x,k_y,z)]^*[-ik_y\hat{\phi}(k_x,k_y,z)]\right\} / \int dz \ , \tag{6}$$

where $\hat{n}$ is the x, y Fourier transform of the density and $\Re(\ldots)$, $(\ldots)^*$ indicates the 'real part' and 'complex conjugation' operations respectively. The QL radial heat flux Q^{QL} is computed exactly the same way as the radial particle flux Γ^{QL}, using Eq. 2, with $\Gamma^L(k_y)$ substituted by the k_y spectral component $Q^L(k_y)$ of the electrostatic part of the linear radial heat flux

$$Q^L = \frac{1}{2}m\langle(M_{20} + M_{02})v_x\rangle = -\frac{1}{2}m\langle(M_{20} + M_{02})\partial_y\phi\rangle \ , \tag{7}$$

where m is the species mass and

$$M_{ij} = \int \delta f v_{\parallel}^i v_{\perp}^j d\mathbf{v} \tag{8}$$

are the ith $v_{\parallel}$ and j^{th} $v_{\perp}$ moments of the fluctuating part δf of the particle distribution function. The choice of the QL weights $(\gamma/\langle k_{\perp}^2\rangle)^{\xi}$ is usually referred as 'mixing length saturation rule', and accounts for the saturation level of the fluctuating potential.

With the same spirit, an 'average' QL estimate of the most unstable mode frequency ω has been defined, according to [13], as a weighted sum over the frequency k_y spectra, with the same QL weights as in Rel. 2:

$$\omega^{QL} = \sum_{k_y}\left(\frac{\gamma_{k_y}}{\langle k_{\perp}^2\rangle(k_y)}\right)^{\xi}\omega_{k_y} / \sum_{k_y}\left(\frac{\gamma_{k_y}}{\langle k_{\perp}^2\rangle(k_y)}\right)^{\xi} \ . \tag{9}$$

This parameter gives a quantitative measure of the relative contribution of different turbulent modes to the QL fluxes. It is positive for a dominant ITG regime, while it is negative for a dominant TEM regime, according to GENE conventions.

2.3. Results in the collisionless regime

As a first step, in figure 3 we plot the k_y spectra of the growth rate γ and the real frequency ω, for the 2 species electron-deuteron case and for the 3 species case including carbon impurity, computed with parameters set to the experimental ones (see Table 1). For each scan we considered 44 k_y values, multiples of $k_{\mathrm{y,min}} = 0.05\rho_s^{-1}$ (the same considered in the NL simulations), where $\rho_s = c_s/\Omega_i$ is the sound Larmor radius, Ω_i being the ion cyclotron frequency. In the three species simulations, carbon has been treated self-consistently as an active species, since the value of Z_{eff} (see Fig. 2 (d)) is close to the critical value $Z_{\mathrm{eff}} \sim 1.6$ identified in [19] for treating impurities as active (i.e. contributing to self-consistent ES field) instead of passive (trace). Nevertheless, as it will be shown, the results indicate that the carbon could have been treated as a trace, at least in the considered collisionless case. The results show that experimental parameters, if we consider the collisionless case, are relative to a TEM dominated microturbulence regime. The 2 species case significantly differs from the 3 species one only for $1 < k_y\rho_s < 1.4$, where an ITG region develops, continuously connected in frequency to a TEM one.

In Fig. 4, the k_y spectra of the particle fluxes for the same case as in Fig. 3 are shown, for the two species and the three species cases. The signs are positive for outward radial flux. The '3 species case' ion particle flux is almost completely superimposed to the '2 species case' one, while

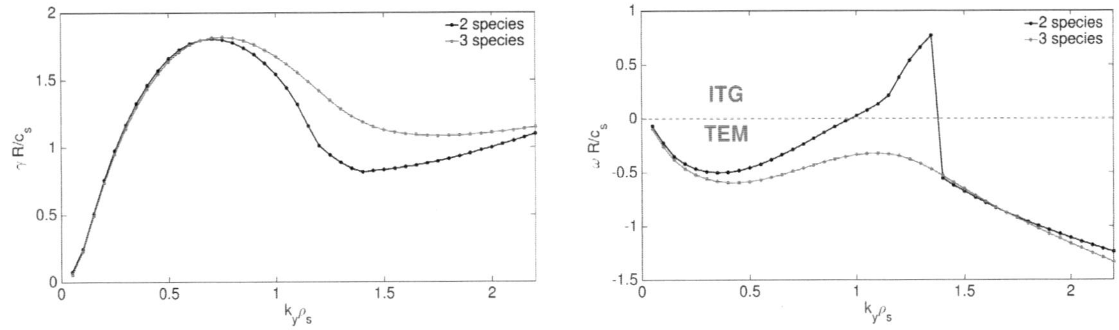

Figure 3. k_y spectra of the growth rate γ and real frequency ω for the most unstable mode, considering only electrons and deuterons (black) and including carbon (red). γ and ω are normalised with respect to c_s/R, while k_y is normalised with respect to $1/\rho_s$.

the electron one is higher than the '2 species case' one, due to the presence of a non-vanishing outward carbon flux. This ensures the ambipolarity condition $\sum_{sp} q_{sp}\Gamma_{sp} = 0$. Nevertheless, the presence of carbon does not modify the (outward) signs of all fluxes. The agreement between the ion particle fluxes in the '2 species case' and in the '3 species' one can be justified noting that, looking at Fig. 3, even if the growth rates are modified in this k_y interval, this is not important in the evaluation of the QL weights since $1/k_\perp^2 \sim 1/k_y^2$ and therefore for $k_y\rho_s > 1$ the behaviour of the QL weights $(\gamma/\langle k_\perp^2\rangle)^\xi$ dominates and zeros the contributions to the fluxes from the $1 < k_y\rho_s < 1.4$ ITG region.

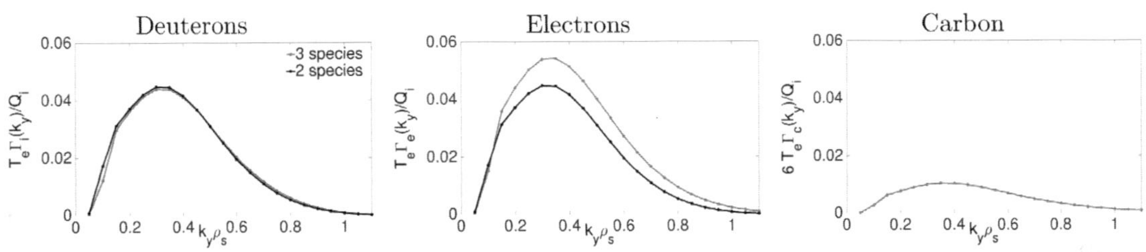

Figure 4. From left to right: k_y spectra of deuteron, electron and carbon normalised particle flux (the particle fluxes are normalised with Q_i/T_e), multiplied by $|q_{sp}|$. The fluxes satisfy ambipolarity. Black indicates the 2 species case, while the 3 species case results are in red. The parameters are set to experimental ones.

For every particle species, the particle flux can be decomposed as follows [13]:

$$\Gamma = \Gamma_0 \sum_{k_y} \left[A_{k_y}\frac{R}{L_n} + B_{k_y}\frac{R}{L_T} + C_{k_y} \right] , \tag{10}$$

where Γ_0 is a constant and A_{k_y}, B_{k_y} and C_{k_y} are the diffusive (diagonal), thermodiffusive and pinch transport coefficients respectively. In general these three k_y-dependent coefficients are themselves functions of the density and temperature gradients of all the active species. Assuming the diagonal transport coefficient A_{k_y} being dominant over the off-diagonal ones (B_{k_y}, C_{k_y}), implies that the particle flux Γ is most sensitive to R/L_n (one should note however that it is not always the case, in particular if the dominant instability is not density gradient driven). in order to find the parameters leading to $\Gamma = 0$, we first consider the 2 species case. We performed three single scans in R/L_n, R/L_{Te} and R/L_{Ti}, varying these parameters in the intervals $R/L_n \in [0, 10]$, $R/L_{Te} \in [0, 16]$ and $R/L_{Ti} \in [0, 16]$, letting the other parameters set to the experimental ones. For each scan, in order to compute the QL fluxes, we further scanned k_y values given by $k_y = n_{k_y} k_{y,\mathrm{min}}$, with $k_{y,\mathrm{min}} = 0.05\rho_s^{-1}$, $n_{k_y} = 1, 2, \dots 28$, in order to well resolve the low k_y part of the spectrum, that is the only one significantly contributing to the flux. Considering only two species, the electron and ion particle fluxes are equal to fulfill ambipolarity, and we will refer to them as $\Gamma \equiv \Gamma_e = \Gamma_i$. The results are plotted in Fig. 5, illustrating that the normalised particle flux is mainly affected by the variation of R/L_n and R/L_{Ti} (as expected). Moreover, note that the flux crosses zero only in the density gradient scan.

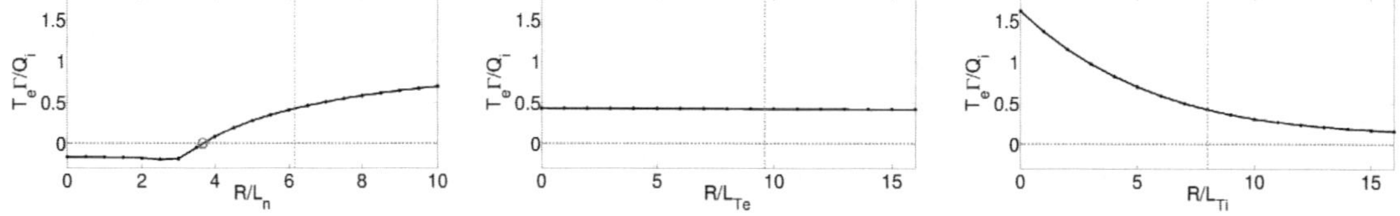

Figure 5. From left to right: particle flux $\Gamma \equiv \Gamma_e = \Gamma_i$, normalised to Q_i/T_e, for R/L_n, R/L_{Te} and R/L_{Ti} scans, considering two species, with other parameters set to the experimental ones. The experimental values of R/L_n, R/L_{Te} and R/L_{Ti} are indicated by vertical pink dashed lines.

In view of the strong dependence of the particle flux on these two parameters, a double scan in the $R/L_{Ti}, R/L_n$ plane has been performed, in order to reconstruct the $\Gamma = 0$ curve. This curve has been obtained performing R/L_n scans with spacing $\Delta(R/L_n) = 0.5$ at fixed values of the ion temperature gradient, in the interval $R/L_{Ti} \in [2, 13]$, with spacing $\Delta(R/L_{Ti}) = 1$. The zero condition has been obtained with linear interpolation of $\Gamma(R/L_n)$. The results are shown in Fig. 6 (a). The $\Gamma = 0$ condition is fulfilled in the $(R/L_{Ti}, R/L_n)$ plane by a non-monotonic function of R/L_{Ti}, that has a maximum at a value of the ion temperature gradient close to the experimental one ($R/L_{Ti,\mathrm{exp}} \simeq 8$). Moreover, the $\omega^{QL} = 0$ curve crosses the $\Gamma = 0$ one close to its maximum, in agreement with [13] (even if in our case the parameters are much different from [13], that considers an eITB case), showing that varying the ion temperature gradient, the maximum value of the electron density peaking factor $R/L_n(\Gamma = 0)$ is obtained in a spectrally balanced TEM-ITG turbulence regime. In particular, the $\Gamma = 0$ curve crosses marginally the experimental lower boundary of the error bar in R/L_n close to its maximum. The flux balancing between TEM and ITG contributions at the maximum is shown in detail in Fig. 6. In Fig. 6 (b) the k_y spectra of the growth rate and frequency of the main unstable mode are presented for the point $(8, 3.68)$ in the $(R/L_{Ti}, R/L_n)$ plane, matching the zero particle flux condition close to the maximum of the $\Gamma = 0$ curve, at experimental R/L_{Ti}. A TEM branch is present for low k_y values, continuously connected to an ITG one, resulting in an average QL frequency ω^{QL} close to zero. In Fig. 6 (c) the k_y spectra of the normalised particle flux and the frequency are compared. It can be seen that the zero particle flux results from the balancing of an outward TEM contribution and an inward ITG one. Finally, in order to show the coexistence of TEM and ITG modes at different wavenumbers and their different balance varying R/L_n, the growth rate γ and the frequency ω of the main unstable mode are shown in Fig. 7 (a) and (b) respectively, versus $k_y \in [0.05, 1.3]\rho_s^{-1}$ and $R/L_n \in [0, 7]$, with the other parameters set to experimental ones. The density gradient $R/L_n = 3.68$, relative to the zero particle flux (Fig.6 (b) and (c)), is indicated by a black dashed line. As it can be seen in Fig. 7 (b), increasing R/L_n a TEM region develops at small wavenumbers, broadening continuously to higher ones. The balanced coexistence in the k_y spectrum of TEM and ITG modes around $R/L_n \sim 4$ allows the fulfilment of the zero particle flux condition.

The identification of the point $(R/L_{Ti} \simeq 8, R/L_n = 3.68)$ close to the zero particle flux curve maximum in the $(R/L_{Ti}, R/L_n)$ plane, constitutes the major result of our 2 species analysis

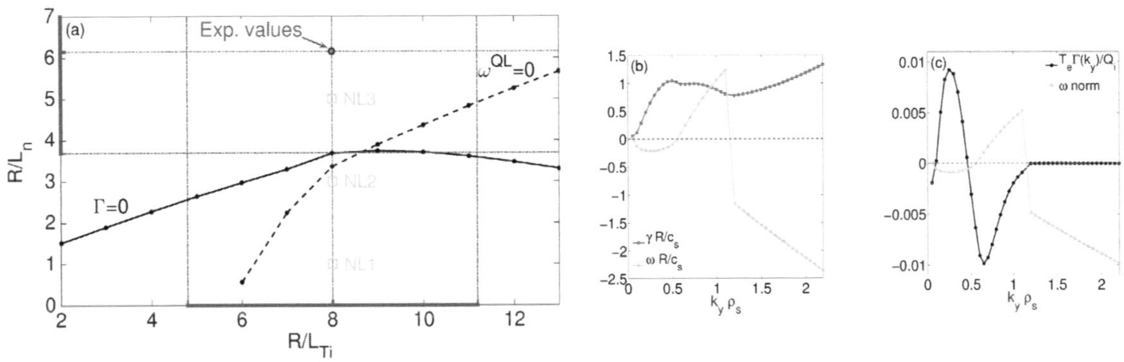

Figure 6. (a) Zero particle flux curve (solid line) and $\omega^{QL} = 0$ curve (dashed line) in the $(R/L_{Ti}, R/L_n)$ plane, considering two species in the collisionless regime. The experimental $R/L_{Ti}, R/L_n$ values and error bars are indicated in blue, while the green squares localize the points chosen for the NL validation of QL results (see section 2.6); (b) k_y spectra of the normalised particle flux and most unstable mode frequency (normalised to the absolute value of the Γ maximum), evaluated at the point $(8, 3.68)$ in the $(R/L_{Ti}, R/L_n)$ plane, matching the $\Gamma = 0$ condition close to the maximum of the $\Gamma = 0$ curve, at experimental $R/L_{Ti} \simeq 8$; (c) k_y spectra of the growth rate γ and real frequency ω, normalised with respect to c_s/R, at the same point in the $(R/L_{Ti}, R/L_n)$ plane.

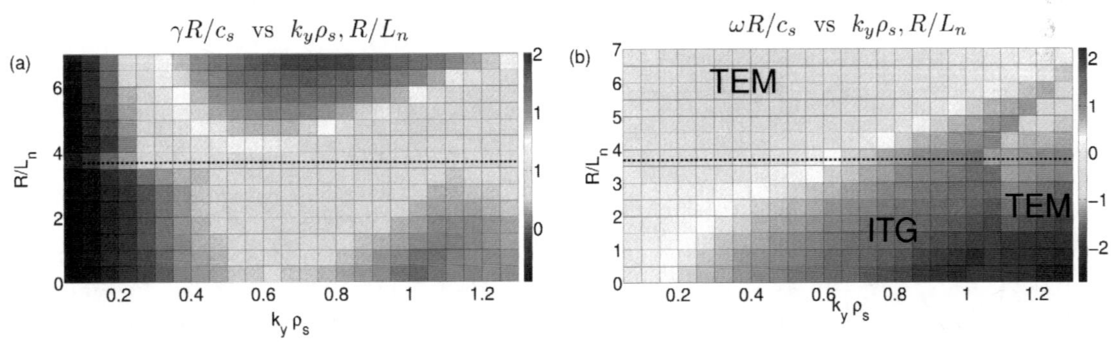

Figure 7. (a) Normalised growth rate $\gamma R/c_s$ for the most unstable mode, plotted versus $k_y\rho_s$ and R/L_n, with the other parameters set to experimental ones. The black dashed line indicates the R/L_n value of Fig. 6 (b) and (c) ; (b) Similar to (a), showing the frequency of the most unstable mode instead of the growth rate.

in the collisionless case, as it confirms the presence of a $\Gamma = 0$ matching set of parameters, marginally within the experimental error bars, consistent with a balanced mixed TEM-ITG regime, consistent with the possible explanation of the rotation reversal as a transition between these two microturbulence modes.

2.4. Effect of carbon impurity on the zero particle flux condition

As a second step in the analysis of the zero particle flux condition, the effects of the carbon impurity have been included. Since the carbon density gradient gives rise to the diffusive (diagonal) part of the carbon particle flux, it is an essential variable to be scanned. A double scan has been performed in the $(L_{ne}/L_{nc}, R/L_{ne})$ plane, with all the other parameters, including

R/L_{Ti}, taken from experimental measurements. Note that this choice of the ion temperature gradient is based on the 2-species results previously discussed. Indeed, in that case the $R/L_{Ti} \simeq 8$ value provided the maximum density peaking factor. For each considered value of R/L_{ne} (y axis), the x axis value L_{ne}/L_{nc} is obtained varying the carbon density normalised logarithmic gradient R/L_{nc}. In order to perform the three species gyrokinetic simulations, for each couple $R/L_{ne}, R/L_{nc}$, the ion density gradient is adapted invoking quasi-neutrality:

$$\frac{R}{L_{ni}} = \frac{1}{n_i}\left(n_e \frac{R}{L_{ne}} - 6n_c \frac{R}{L_{nc}}\right). \tag{11}$$

In all the simulations, carbon has been treated as a self-consistent (active) species. The three curves relative to $\Gamma_e = 0$, $\Gamma_i = 0$ and $\Gamma_c = 0$ have been obtained in the $(L_{ne}/L_{nc}, R/L_{ne})$ plane, aiming to find a common intersection point to be identified as the common zero particle flux point (note that if two out of three curves cross at one L_{ne}/L_{nc} value, the third has to cross at the same value due to ambipolarity). The zero particle flux matching values of the electron density gradients are computed for each species by linear interpolation of flux profiles obtained scanning R/L_{ne} with spacing $\Delta(R/L_{ne}) = 1$ for fixed $L_{ne}/L_{nc} = (1/3, 1/2, 3/4, 1, 4/3)$ values of the ratio between carbon and electron logarithmic density gradients. The results are shown in Fig. 8. A common zero particle flux point $\{\Gamma_{sp} = 0\ ,\ sp = e, i, c\}$ is found, ad it is located at $(L_{ne}/L_{nc} \sim 0.49, R/L_{ne} \sim 3.56)$, corresponding to $R/L_{ni} \sim 3.79$ and $R/L_{nc} \sim 1.74$, since $n_i/n_e \sim 0.89$ and $n_c/n_e \sim 0.02$ at $\rho_{\mathrm{tor,n}} = 0.6$. The $R/L_{ne}(\Gamma = 0)$ is within 3% agreement with the two species result, therefore the effect of carbon on the electron density peaking factor $R/L_n(\Gamma = 0)$ appears to be relatively minor in the collisionless limit.

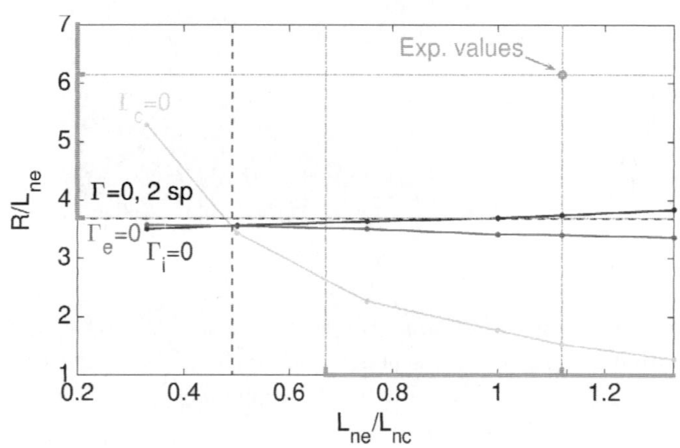

Figure 8. $\Gamma_e = 0$ (red), $\Gamma_i = 0$ (dark blue), $\Gamma_c = 0$ (light blue) curves in the $(L_{ne}/L_{nc}, R/L_{ne})$ plane considering three species. For comparison, the $R/L_n(\Gamma = 0)$ value of the two particle case is plotted as a dashed horizontal black line. The experimental values and error bars of L_{ne}/L_{nc} and R/L_{ne} are indicated in pink.

2.5. Preliminary results for the effect of collisions on the $R/L_n(\Gamma = 0)$ determination

In this section some preliminary results on the effect of collisions on the determination of the zero particle flux matching density gradient are presented. To start, considering for simplicity only two species, the k_y spectra of the growth rate and frequency have been computed taking into account collisions, for parameters set to the experimental ones. The results, shown in Fig. 9, where these spectra are compared with the ones in the collisionless limit, indicate that collisions, as expected, have a damping effect on TEM modes. Moreover, the collisions result in a strong decrease of the growth rate, with almost a factor 2 on the peak value of the spectrum. To investigate the effects of the collisions on $R/L_n(\Gamma = 0)$, a double scan in R/L_n and normalised collisionality $\bar{\nu}$ has been performed, with $R/L_n \in [0, 7]$, $\Delta(R/L_n) = 0.5$

Journal of Physics: Conference Series **775** (2016) 012007 doi:10.1088/1742-6596/775/1/012007

and $\bar{\nu}/\bar{\nu}_{exp} = (0, 1/3, 2/3, 1)$, $\bar{\nu}_{exp} = 1.26$, considering two species and other parameters set to experimental ones. The result of this analysis is shown in Fig. 10. Here, the normalised particle flux is plotted versus R/L_n for the different values of the collisionality. One notices that $R/L_n(\Gamma = 0)$ decreases with increasing collisionality, while $R/L_n(\omega^{QL} = 0)$ increases. These results are in very good agreement with [20]. The phenomenon can be explained as follows: due to collisions, TEM modes are damped and, at the same R/L_n, ITG contributions to the spectrum tend to become more important, shifting upwards the value of $R/L_n(\omega^{QL} = 0)$. Then, in the resulting ITG dominated regime, collisions produce a net outward particle flux that develops firstly at the lowest wavenumbers, spreading continuously to higher ones with increasing collisionality, thus shifting the $R/L_n(\Gamma = 0)$ to lower values. As further explained in Ref. [20], this outward ITG particle flux results from the different interplay between low energy and high energy trapped electrons (associated to inward and outward flux respectively) in the collisionless and collisional regimes.

The main conclusion of this analysis is that adding the effect of collisions, the $\Gamma = 0$ matching density gradient is lowered, falling definitely out of the experimental error bars. However, other relevant parameters such as R/L_{Te} and T_e/T_i have to be varied within the error bars, and it is expected that they could have an impact on the zero particle flux determination, since usually TEM drive increases at higher R/L_{Te} and higher T_e/T_i, resulting in a different turbulence regime, leading to different linear fluxes and QL weights. It therefore remains to be checked if their effect could compensate or even overcome the one of collisions.

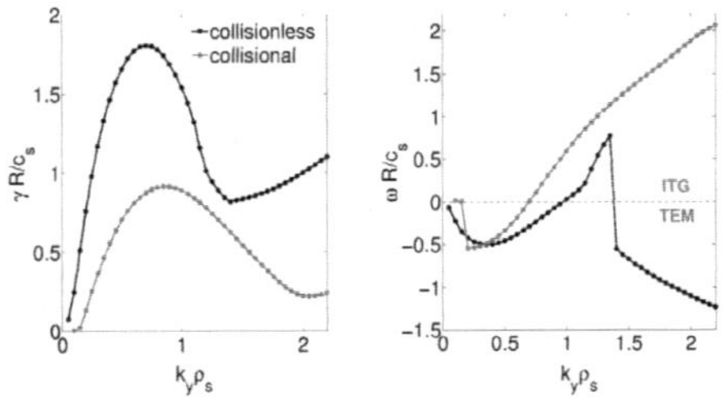

Figure 9. k_y spectra of the growth rate and frequency for the collisionless (black) and collisional (red) regimes, considering two species, with parameters set to experimental ones.

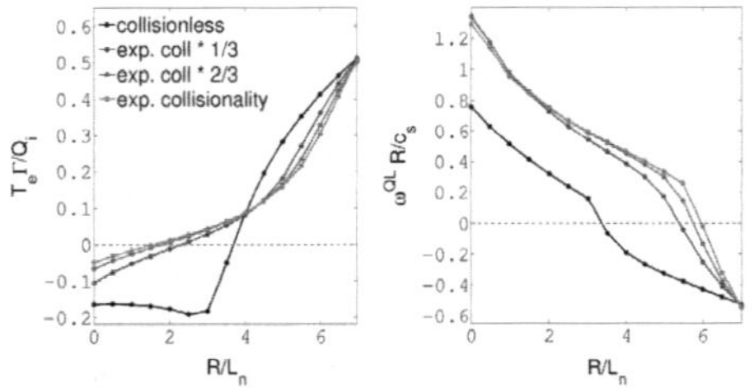

Figure 10. (left) normalised particle flux versus normalised logarithmic density gradient, varying collisionality from zero to the experimental value; (right) ω^{QL} versus R/L_n. ω^{QL} is normalised with respect to c_s/R.

2.6. Non-Linear validation of the Quasi-Linear results

In order to validate the QL results, NL simulations have been performed for a reduced number of cases, comparing the particle fluxes with the QL ones. In Fig. 11, the NL fluxes of the simulations $NL1, NL2, NL3$ indicated by the green squares in Fig. 6 are compared with the QL estimates (R/L_n scan, $R/L_n \in [0, 7]$, $\Delta(R/L_n) = 0.5$, all other parameters = exp. values) computed with the three choices $\xi = 1, 2, 3$ of the QL weights exponent. Here, the two species collisionless case is considered for simplicity. For the NL simulations, the values $R/L_n = 1, 3$ and 5 were considered, and they have been performed considering a $n_{kx} \times n_{ky} \times n_z \times n_{v\|} \times n_\mu = 256 \times 64 \times 32 \times 64 \times 16$ grid. As can be inferred from these results, the agreement in the QL estimates between the different choices of ξ is good and the agreement between NL and QL estimates is also satisfactory, even if it is better at higher density gradients. In particular, the best agreement between QL and NL results is with the choice $\xi = 3$, but one can note that the results obtained with $\xi = 2$ are not significantly different. Therefore, our initial $\xi = 2$ guess (following [13]) is justified. The overall NL behaviour, i.e. the particle flux going from negative to positive increasing density gradient, is well reproduced by the QL estimates. Despite this, the NL value of $R/L_n(\Gamma = 0)$ is lowered with respect to the QL estimates, even further away from the experimental density gradient. In order to investigate the behaviour of the NL results at lower R/L_n, more NL simulations still have to be performed.

In Fig. 12 the NL and QL ($\xi = 2$) k_y spectra of particle fluxes for the $NL1$ and $NL3$ simulations are compared. The QL spectra give a relatively good approximation of the NL ones. Differences may be attributed to the fact that in the QL model only the main unstable mode at each k_y is taken into account, while in NL simulation there are always different modes contributing, and in general ITG and TEM branches coexist at a same k_y [21].

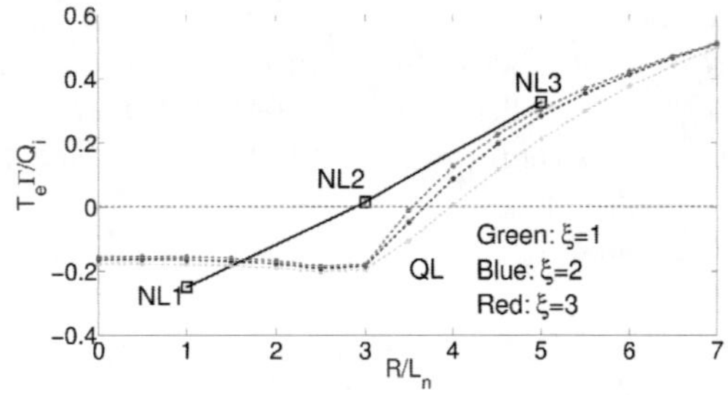

Figure 11. NL-QL comparison of the normalised particle fluxes. Non-Linear simulations have been carried out for cases NL1, NL2, NL3 corresponding to the green points in Fig. 6, with $R/L_n = 1, 3$ and 5, with all other parameters set to experimental ones. These simulations are performed with two species, in the collisionless regime. The QL exponents was in turn set to $\xi = 1, 2, 3$. NL results are shown by a solid black line, while QL results are plotted in dashed coloured lines.

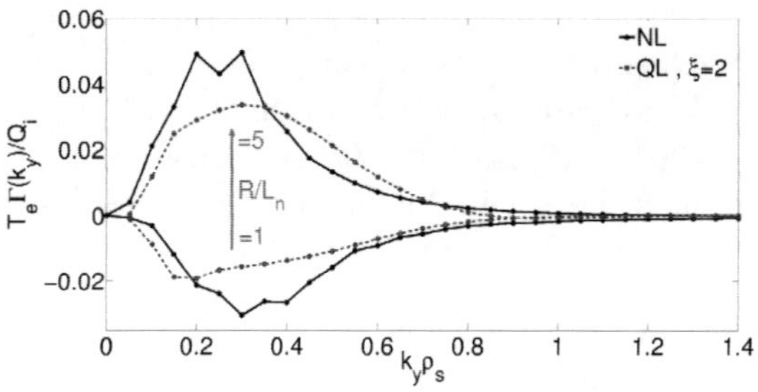

Figure 12. NL-QL comparison of the k_y spectra of the normalised particle flux for the $NL1$ and $NL3$ simulations ($R/L_n = 1$ and 3 respectively).

3. Conclusions

Considering the TCV tokamak shot # 28355, showing toroidal velocity reversal during a density ramp up, we have presented a study of the plasma parameters matching the zero particle flux condition close to the rotation reversal, aiming to address the possible explanation of the reversal as a consequence of a microturbulence transition from a TEM dominated regime to an ITG dominated one. This study has been carried out by means of gyrokinetic local flux-tube simulations with the GENE code, adopting a quasi-linear model to compute the fluxes, and validating the results with non-linear simulations for a subset of cases. Multiple scans in the main plasma parameters that influence the particle fluxes have been done, comparing the results of the two species deuteron-electron case with the ones obtained including the effects of the main plasma impurity, that is carbon. So far, the major part of this study has been carried out in the collisionless regime, but some preliminary results are shown to address the effects of collisions.

In the collisionless regime, performing a $R/L_{Ti}, R/L_n$ double scan (with other parameters set to experimental ones) with two species, we found a set of parameters that matches the zero particle flux condition within the experimental error bars, corresponding to a quasi-linearly balanced ITG-TEM regime, in agreement with the usual rotation reversal explanation. However this would be expected at low and high densities, since one has no central particle source in all cases during the scan. The next step is to compare with the low and high density cases, and taking into account collisions as it shifts the ω^{QL} value towards ITG.

Doing a $L_{ne}/L_{nc}, R/L_{ne}$ double scan including carbon, with other parameters set to experimental values, a common zero particle flux point $\{\Gamma_{sp} = 0\ ,\ sp = e, i, c\}$ has been found, showing that the three species case result does not differ significantly from the two species one.

The preliminary results obtained including the effects of collisions in the two species case indicate that the density gradient matching the zero particle flux condition in the collisionless regime is lower when the collisionality is non-zero, and therefore the collisional results fall outside the experimental error bars. Nevertheless, other important parameters such as the electron temperature gradient R/L_{Te} and the electron to ion temperature ratio T_e/T_i have to be scanned as well to have a complete picture of the collisional case, and the effect of the carbon impurity also has to be accounted for.

Finally, since TCV is a small-sized tokamak, it is expected that finite $\rho^\star$ effects could lead to a significant discrepancy between local flux-tube results and global ones [22]. Previous turbulent transport studies of TCV discharges using flux-tube models, addressing internal transport barriers [14] and plasma shaping effects [23] have pointed out the importance of these effects. This will involve carrying out simulations using the global version of the GENE code.

References

[1] Burrell K H 1997, *Phys. Plasmas* **4**, 1499
Terry P W 2000, *Rev. Mod. Phys.* **72**, 109
[2] Wolf R C 2003, *Plasma Phys. Control. Fusion* **45**, R1-R91
[3] Zastrow K D *et al.* 1998, *Nucl. Fusion* **38**, Number 2
[4] Hutchinson I H *et al.* 2000, *Phys. Rev. Lett.* **84**, 3330
[5] Special issue on ITER Physics Basis [*Nucl. Fusion* **39**, 2175 (1999)]
[6] Hofmann F *et al.* 1994, *Plasma Phys. Control. Fusion* **36**, B277.
Rice J E *et al.* 2007, *Nucl.Fusion* **47**, 1618
[7] Bortolon A 2006, *et al.*, *Phys. Rev. Lett.* **97** 235003
[8] Duval B P *et al.* 2007, *Plasma Phys. Control. Fusion* **49** B195-B209
[9] Diamond P H *et al.* 2013, *Nucl. Fusion* **53**, 104019
[10] Duval B P *et al.* 2008, *Phys. Plasmas* **15** , 056113
[11] Peeters A G *et al.* 2011, *Nucl. Fusion* **51**, 094027
[12] Parra F I and Barnes M 2015, *Plasma Phys. Control. Fusion* **57**, 045002
[13] Fable E *et al.* 2010, *Plasma Phys. Control. Fusion* **52**, 015007
Dannert T and Jenko F 2005, *Physics of Plasmas* **12**, 072309
Jenko F *et al.* 2005, *Plasma Physics and Controlled Fusion* **47**, B195
[14] Lapillonne X *et al.* 2011, *Plasma Phys. Control. Fusion* **53**, 054011
[15] Jenko F *et al.* 2000, *Phys. Plasmas* **7**, 1904
Görler T *et al.* 2012, *J. Comput. Phys.* **230**, 7053
[16] Luetjens H *et al.* 1999, *Computer Physics Communications* **97**, 219
[17] O. Sauter *et al.* 2014, *Phys. Plasmas* **21**, 055906
[18] Jenko F and Scott B D 1999, *Phys. Plasmas* **6**, 2418
[19] Fable E, PhD Thesis, Thesis Number 4334, CRPP Lausanne (2009).
[20] Angioni C *et al.* 2005, *Phys. Plasmas* **12**, 112310
Angioni C *et al.* 2009, *Phys. Plasmas* **16**, 060702
[21] Merz F and Jenko F 2010, *Nucl. Fusion* **50**, 054005
[22] McMillan B F *et al.* 2010, *Phys. Rev. Lett.* **105**, 155001
[23] Merlo G *et al.* 2015, *Plasma Phys. Control. Fusion* **57**, 054010

Journal of Physics: Conference Series **775** (2016) 012008 doi:10.1088/1742-6596/775/1/012008

Correlations and non-local transport in a critical-gradient fluctuation model

J H Nicolau[1], L García[1] and B A Carreras[2]

[1] Universidad Carlos III, 28911 Leganés, Madrid, Spain

[2] BACV Solutions, 110 Mohawk Road, Oak Ridge, Tennessee 37830, USA

E-mail: javherna@fis.uc3m.es

Abstract. A one-dimensional model based on critical-gradient fluctuation dynamics is used to study turbulent transport in magnetically confined plasmas. The model exhibits the self-organized criticality (SOC) dynamics. At the steady state, two regions are found: the outer one is close to critical state and the inner one remaining at the subcritical gradient. The gradient-flux relation exhibits a parabola-like profile centered in the most probable gradient following experimental studies. This is a signature of the non-locality of particle transport driven by avalanches: at the given position transport is due to gradients situated into closer but different positions. The R/S analysis, applied to the fluxes dynamics reveals memory and correlation. Different H exponents corresponding to different dynamical behavior are obtained. The flux at the edge exhibits long time correlations, which can be suppressed if the external drive or the system size is modified. On the other hand, we found that in the sub-critical region the quasiperiodicity is present in the avalanches.

1. Introduction

Since several decades transport in magnetically confined plasmas has been an important topic of research. Self Organized Criticality (SOC) dynamics has been introduced in the field of complex systems [1] and then used as a simple paradigm for turbulent transport studies in magnetically confined plasmas [2] [3]. The SOC systems have to be seen as a qualitative method for studying basic physical behavior. They exhibit some of the characteristics of the L-mode such as profile stiffness, Bohm scaling and superdiffusion [4], which are out of scope in the framework of the diffusive model.

Here we consider a one-dimensional transport model based on the critical-gradient fluctuations dynamics. This model was presented in Ref. [5]. It has the properties of a SOC system. However, the transport is represented by a continuous amount of particles instead of an integer amount. Within this approach, we make an analogy from the height h of a sandpile to the average particle density in the plasma.

In the present work we are studying some of the characteristics of this model as the non-diffusive transport, non-locality and long time correlations.

Recently, experimental data from different devices [6] indicated that the relation between gradient and radial flux is not linear as in a diffusive model. Our simple transport model exhibits the same behaviour. The gradient-flux relation is given by a parabola-like profile centered in the most probable gradient. This indicates the non-locality of avalanches as an underlying transport mechanism.

The long-time correlations have been observed in the model as expected in a SOC system. Through the R/S analysis [7] [8] the quasi-periodicities of the model, which were described in Ref. [9] are confirmed.

In addition, in magnetic nuclear fusion devices, the protection of the wall is an important issue. It must support high temperatures and heat pulses at long time scales. That is why the fluxes should be studied as global events rather than an instantaneous process. In the Sec.5 we will study some characteristics of flux pulses produced by avalanches in our model.

This paper is organised as follows: in the Sec. 2 the one-dimensional transport model is introduced. The Sec. 3 is dedicated to investigation of dependency between gradients and fluxes. The Sec. 4 deals with the studies of transport correlations. Flux pulse are considered in the Sec. 5. Finally, comments and conclusions are discussed in the Sec. 6.

2. Model description

The transport of the averaged density $h(x)$ is driven by the root-mean-square fluctuations $\phi(x)$ as follows

$$\frac{\partial h}{\partial t} = \frac{\partial}{\partial x}\left(\mu_0 \phi \frac{\partial h}{\partial x}\right) + S_0, \tag{1}$$

$$\frac{\partial \phi}{\partial t} = \phi\left(\gamma - \mu\phi\right) + S_1. \tag{2}$$

The Eq. 1 is a transport equation containing the radial diffusion term and the source term S_0. Here the diffusivity is proportional to $\mu_0\phi$ where ϕ are the fluctuations driven by the Eq. 2 and μ_0 is a constant term to regulate the diffusivity. The latter is called the evolution equation. It contains a linear term responsible for the linear triggering of the instability with growth rate γ, a nonlinear term with a coefficient μ, which brings the fluctuations to the saturation level and a source S_1, which guarantees a minimum level of fluctuations in the numerical simulations. The coefficient γ is proportional to the critical gradient of the instability:

$$\gamma = \gamma_0\left(-\frac{\partial h}{\partial x} - z_c\right)\Theta\left(-\frac{\partial h}{\partial x} - z_c\right).$$

Here z_c is the value of the critical gradient and Θ is the Heaviside step function. Therefore, if the gradient $Z = -\partial h/\partial x$ is lower than the critical gradient then the linear growth γ is equal to zero and the fluctuations decrease. In the same way, if the gradient is higher than z_c, it triggers the fluctuations and therefore the radial diffusion in the Eq. 1 increases.

This transport model is an analogy of the classical running sandpile. The linear term in the Eq. 2 proportional to γ plays the role of the *threshold* in the sandpile, the ratio of coefficients μ/μ_0 regulates the *drive* of the system and S_0 can be interpreted as the new grains of sand we add to the system.

The equations are numerically advanced as follows. We add a small quantity to the source terms with some probability, that is

$$\phi_i^t \to \phi_i^t + \bar{\phi}$$

with probability p_1, and

$$h_i^t \to h_i^t + \delta$$

with probability p_0.

The transport equation given by the Eq. 1 and the evolution equation given by the Eq. 2 are numerically implemented with using the following time-discretisation scheme:

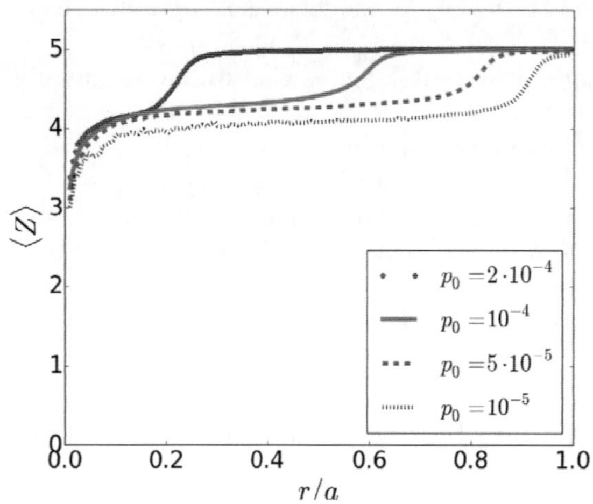

Figure 1. The average slope of the profile $\langle Z \rangle$ is plotted for different values of p_0 (probability to add new grains). Two regions are formed with a clear transition between them. Outer region remains very close to the critical gradient value ($z=5$) and the inner one at the subcritical state ($z<5$). L=200.

$$h_i^{t+\Delta t} = h_i^t + \Delta t \mu_0 \left(-\phi_i^{t+\Delta t} z_i + \phi_{i-1}^{t+\Delta t} z_{i-1} \right),$$

$$\phi_i^{t+\Delta t} = \phi_i^t \exp\left[\Delta t \left(\gamma_i - \mu \phi_i^t \right) \right],$$

where $\gamma_i = (z_i - z_c)\,\Theta(z_i - z_c)$, with $z_i = h_i^t - h_{i+1}^t$.

We are using the Dirichlet boundary condition $h\,(L) = 0$ where L is the size of the system. For all the results included in this paper the parameters of the simulations have the following values: $z_c{=}5$, $\gamma_0{=}1$, $\mu{=}200$, $\mu_0{=}100$, $\Delta t{=}0.05$, $\bar{\phi}{=}10^{-8}$, $p_1{=}10^{-7}$ and $\delta{=}0.05$.

The numerical results show that the steady state can be reached. We will focus on the slope Z of the sandpile height h. On the Fig. 1 the average slope is presented. We observe that the system self organizes in two regions as follows: the inner region remains at the subcritical state ($z{<}z_c$), which is a characteristic of SOC systems, while at the outer region a critical region ($z{=}z_c$) is formed.

In the subcritical region, transport is mostly avalanche-like. Furthermore it exhibits a quasiperiodicity [9] similar to the sandpile model with diffusion [10]. In the critical region the transport is quasicontinuous as we will see in the Sec. 4. The transition point (normalized to L) between zones is proportional to $p_0 L^{3/4} \left(\mu/\mu_0\right)^{7/8}$ [5]. In the present paper we will only vary the parameter p_0 which represents the probability to add a sand grain inside the sandpile. We will use $L = 200$ or $L = 1200$ to highlight some of the characteristics of system.

3. Relation between gradient and flux

Let us focus on the gradients Z and fluxes $\Gamma = \mu_0 \phi Z$ inside the system. In any diffusive model the flux is proportional to the gradient. When the gradient increases the flux rises as well. On the same way, lower gradients implies lower fluxes. However, in [6], fluxes and gradients were measured at the edge of different magnetically confined fusion devices: not only the proportional behavior has been not observed but also higher fluxes were found for low gradients. Averaged measurements have shown that the gradients below the mean value involve higher fluxes.

Our simple model exhibits a similar behavior as can be seen on the Fig. 2. On the abscissa we put the values of the gradient normalized to its standard deviation σ. The positive values represent steeper than mean value of the gradient and the negative ones: lower than the mean value of the gradient correspondingly. On the ordinate, we put the values of the flux normalized

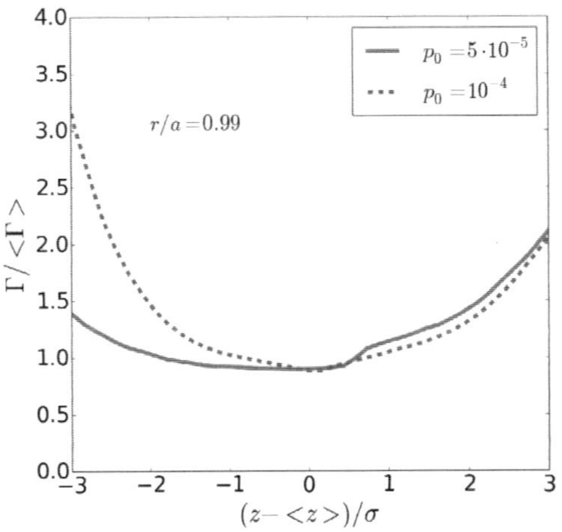

Figure 2. The gradient-flux relation for two different values of p_0. On the abscissa axis: values of gradient normalized to its standard deviation σ. On the ordinate axis: values of flux normalized to its mean value. L=200.

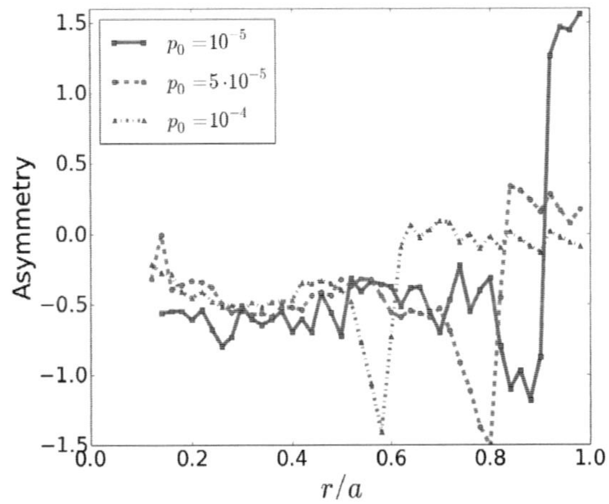

Figure 3. The asymmetry function of the gradient-flux relation as a function of position for three different p_0. Two symmetry regions can be observed for each value of p_0. L=200

to its mean value. The examples on the Fig. 2 represent measurements at the edge of the system following the experimental setup.The parabolic profile is observed, which supports the idea from the experiments: the system is not following a diffusive transport. This result is observed as well at any other position and for different values of p_0, which corresponds to a different size of the subcritical region.

Because the gradient-flux relation follows a parabolic profile centered in the most probable gradient, it is interesting to study the symmetry of that profile and how it behaves accordingly to the different parameters of our simulations. We define the asymmetry function as follows:

$$\text{Asymmetry} = \frac{\int \text{flux} \times \text{gradient}}{\int \text{flux}}.$$

where fluxes and gradients are normalized following the results on the Fig. 2. The asymmetry function gives us zero if the profile is completely symmetric (a parabola), positive values of the right hand size of the profile presented at the Fig. 2 reaches higher values with respect to its left hand side and the negative ones if inversely, the left hand side of the profile reaches higher values.

The symmetry of the gradient-flux relation for different p_0 is presented on the Fig. 3 as a function of the radial position. A discontinuity is observed in each studied case. This is due to the fact that the position of the measurements reaches the transition between the two regions (subcritical and critical). Furthermore, the symmetry has a dramatic change if the radial position is situated inside the one region or inside the other one. If the radial position where we take a measurement belongs to the subcritical region then the asymmetry will be negative. However, for positions situated in the critical region the asymmetry is close to zero or positive. In addition, for positions close to the edge the asymmetry increases as p_0 decreases (the latter leads to the growth of the subcritical region).

Journal of Physics: Conference Series **775** (2016) 012008 doi:10.1088/1742-6596/775/1/012008

However, the value of zero asymmetry for cases of a large critical region does not mean that the gradient-flux relation is giving always a perfect parabola. In fact, for $p_0 > 5 \cdot 10^4$ the flux is practically constant and the profile is no longer a parabola but a flat straight line.

Notice, that this simple one-dimensional transport model reproduces the transport behavior observed experimentally [6]. The physical properties of our model are similar to the sandpile: transport is driven through avalanches. We suggest to refer the parabolic behavior of the gradient-flux relation as a non-local transport effect. We suppose that this is due to the fact that the avalanches at close positions can trigger fluxes. For example, a steep gradient situated in an inner position generates a high flux (mass movement), which can be measured at outer positions even with low gradients.

4. Correlations studies. The R/S analysis

In SOC systems, avalanches are produced by the history of the profile, which characterises the dynamical system. In other words, it means that the presence of avalanches is strongly related to presence of memory and correlation [11]. We have studied the correlations in the flux Γ through the R/S analysis [7] and the calculation of the Hurst exponent [8]. From a time series $x = \{x_1, x_2, \ldots, x_n\}$ of n values, then the rescaled range R/S is defined as:

$$\frac{R(n)}{S(n)} = \frac{\max\left(0, W_1, W_2, \ldots, W_n\right) - \min\left(0, W_1, W_2, \ldots, W_n\right)}{S(n)},$$

where $W_k = x_1 + x_2 + \ldots + x_k - k\bar{x}$, $\bar{x}$ is the average value of time series and S is the standard deviation. If the signal is self-similar, then $R(n)\,/S(n) \sim n^H$, where H is the Hurst exponent. A value $H = 0.5$ indicates a random series. For values higher than 0.5 the data exhibits correlation and memory. However, values lower than 0.5 illustrate anti-correlation. In a SOC system the Hurst exponent is close to $H \sim 0.8$ [11].

On the Fig. 4 and the Fig. 5 the results of the R/S analysis for the flux at the different positions inside the critical and the subcritical regions are respectively shown. We observe that there are two common domains, we call them A and C, on both figures.

The domain A exhibits a Hurst exponent close to the value of 1, which means the presence in the system of a strong correlation. The width of the domains A gives us an estimation of the correlation time. The latter can be modified if the ratio between sizes of critical/subcritical regions changes: wider subcritical region corresponds to longer correlation time and wider A domain. The strong correlation of the dynamics comes from the fact that once the avalanche is started, it is most probable that the avalanche continues on the next time step.

Another common domain is the domain C. The presence of the anti-correlated exponent can be explained due to the finite size of our model. The following behavior is observed for a very long period of time. At some moment all the cells (space discretization) are close to the critical value and a huge avalanche takes place, which moves the system far below the critical point and for some time no avalanche takes place until the system is filled again with new grains. Furthermore, we have observed how the edge between the domain C and the previous one moves in function of the system size variation. The domain A and C have been also observed in SOC systems [11].

The domain B is specific for the subcritical zone on the Fig. 4. For stationary, non-periodical data the H exponent varies from 0 to 1. We have measured a higher value preceded by a value close to zero because the system has quasi-periodicities. Peaks are found in the power spectra (not shown here) that indicates the periodicity of avalanches in the system . Nevertheless, they were observed and discussed in a previous work [9]. That behavior is the same as the observed in the sandpile with diffusion [10]. The size of this domain increases as the system increases because the size of the domain C decreases.

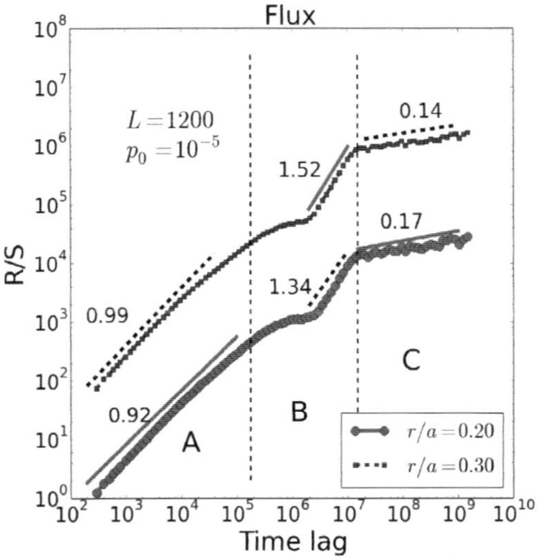

Figure 4. The R/S analysis at the subcritical positions. The ordinate axis values haven been shifted for convenience. Three domains with different dynamics are observed. The domain A demonstrates the correlations of avalanches with themselves. The domain B illustrates the quasi-periodicity. The finite size effects belong to the C domain.

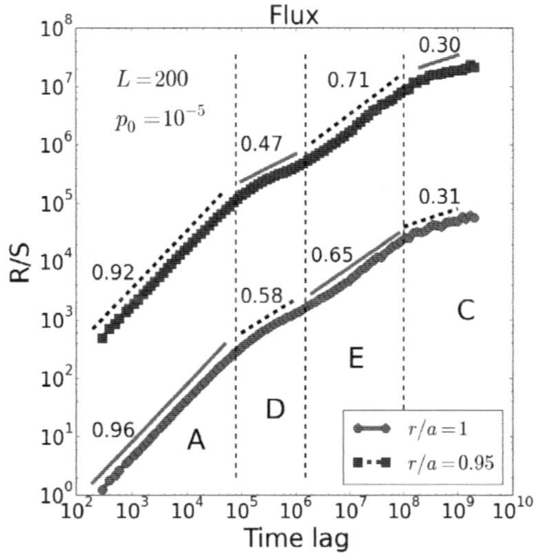

Figure 5. The R/S analysis at the critical positions. The ordinate axis values haven been shifted for convenience. Two additional domains to the Fig. 4 are observed. The appearance of the domain D is due the proximity to the critical state. The long time correlations are present in the domain E.

For the critical region on the Fig. 5, we observe a non correlated or random domain in the domain D. The randomness of the $H \sim 0.5$ is due the fact that the system is at the critical state ($z = z_c$) at that position. A random $H \sim 0.5$ is a characteristic of diffusive systems. In addition, for simulations where the critical region is dominant or L is big enough and the critical region is wider, the E domain can disappear and is replaced by the D one.

Finally, the E domain exhibits correlations at long times for the flux. It is one of the main characteristics of the SOC systems [11], it states that our system has memory for long times. Even at the critical region the model exhibits no diffusive transport and that is why the gradient-flux relation still exhibits the parabola-like profile at the edge. Flux created by the avalanches in the interior can travel to the critical region and modify the transport.

5. Flux pulses

In the subcriticial region the flux is bursty as demonstrated on the Fig. 6, however at the outer region there is a continuous flux with perturbations, which leads to the long time correlations. The flux data exhibits a sawtooth profile with very fast growths and long tails during the relaxation. Values of flux vary between several orders of magnitude. Remark that the avalanches in the model can propagate inward [5] while fluxes are always positive, pointing outward.

Since the flux in the subcritical region exhibits an intermittent behavior, we will not focus on the instantaneous flux but on the integrated flux (or pulses). We have used two kind of methods to measure the pulses. The first one integrates the flux from a local minimum to the next one. The second method demeans the flux data and then the pulse is defined from a zero value to the next zero value (but only positive ones). Therefore the last method eliminates all the small

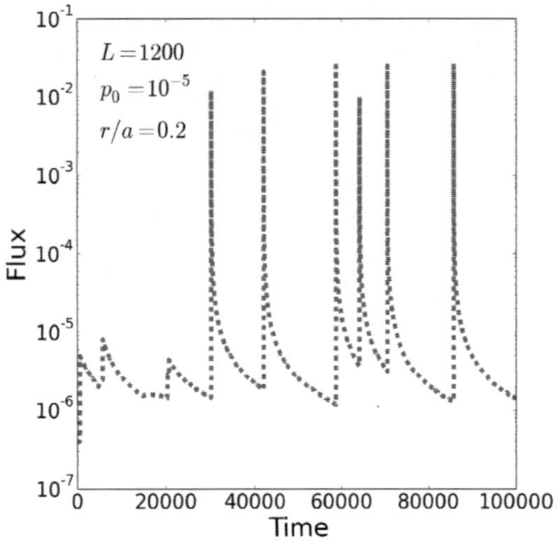

Figure 6. Flux at a subcritical position.

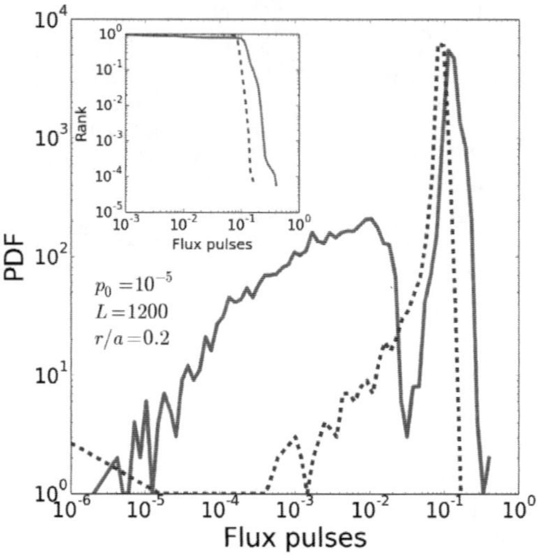

Figure 7. The PDF of the pulses (solid line) and the demeaned ones (dashed line). Inset figure: Rank function.

pulses.

The PDF of both methods is presented on the 7. The solid line corresponds to the first method: we can observe a hump and a sharp peak towards local maximum. The second one (dashed line) gives only one global maximum. Therefore we can conclude that the presence of the sharp peak is a consequence of the large pulses because the second method lacks of the small ones.

The peak in the PDF indicates that most of the larger pulses have similar size. In fact, the rank function (inset figure in Fig. 7) reinforces that statement.

Furthermore, we studied the correlations of the flux pulses. On the Fig. 8 the R/S analysis for two different positions at subcritical and critical regions is presented. The subcritical position exhibit three domains. The first domain reveals a correlation among pulses. Then, a higher value of H is identified, which agree with the quasiperiodicity of the system shown in the domain B. The finite system size effect observed in the domain C are also observed on long times. On the other hand, for positions close to the edge, a correlation is observed again. Then, as the system is larger in this example ($L = 1200$) the domain D dominates and the pulses behave as a diffusive system. Due to the randomness, no system size effect is observed.

6. Conclusions

The one-dimensional model based on critical-gradient fluctuations has been used to reproduce the experimental results of the Ref. [6]. The model shows that a simple dynamical system with avalanches exhibits similar transport properties. The avalanches can generate flux in different positions in a way that the relation between gradient and flux exhibits parabola-like profile, which in its turn indicates the non-locality of transport.

In addition, we have studied through the R/S analysis the correlations in the system. Within that analysis, the quasi-periodicities have been identified inside the subcritical region, the similar behavior has also been observed in Ref. [9]. Then, we have measured long time correlations, which have been studied in SOC systems [11]. It indicates that the considered one dimensional transport model has a long time memory.

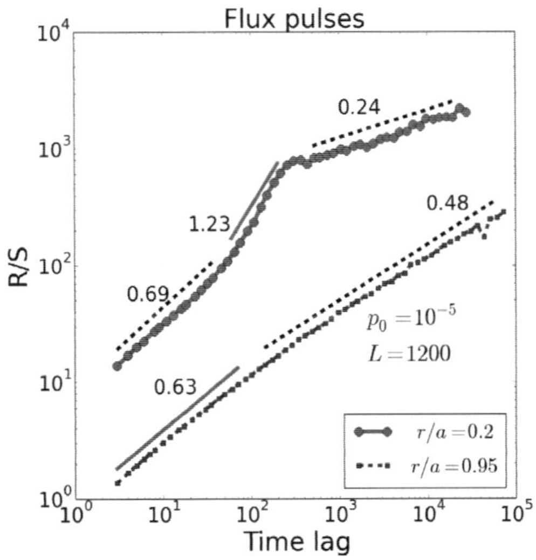

Figure 8. The R/S analysis of the flux pulses at positions in the critical and subcritical region. The inner position exhibits the correlation between pulses, then the quasiperiodicity and the anticorrelation due to finite system size. The edge position exhibits the low correlation followed by random behavior.

Finally, we have examined some of the characteristics of the flux pulses in the system. The PDF demostrates that the majority of pulses are the largest ones and have a similar size. The R/S analysis reveals that the flux pulses are also correlated and the quasiperiodicity of dynamics is present.

7. Acknowledgments

Authors gratefully acknowledge very useful discussion with D. Newman and R. Sánchez. This research is sponsored by Ministerio de Economia y Competitividad of Spain under project ENE2012-38620-C02-02 and UNC313-4E-2361. Fruitful interactions with members of the ABIGMAP research network, funded by the Spanish National Project MAT2015-69777-REDT, is also acknowledged.

8. References

[1] Bak P, Tang C and Wiesenfeld K 1987 *Phys. Rev. Lett.* **59** 381-384
[2] Diamond P H and Hahm T S 1995 *Phys. Plasmas* **2** 3640
[3] Newman D E, Carreras B A, Diamond P H and Hahm T S 1996 *Phys. Plasmas* **3** 1858
[4] Sánchez R, Newman D E 2015 *Phys. Control Fusion* **57** 123002
[5] García L, Carreras B A and Newman D E 2002 *Phys. Plasmas* **9** 841
[6] Hidalgo C, Silva C, Carreras B.A., van Milligen B Ph, Figueiredo H, Garcia L, Pedrosa M A, Gonçalves B and Alonso A 2012 *Phys. Rev. Lett.* **108** 065001
[7] Mandelbrot B and Wallis J 1969 *Water Resour. Res.* **5** 967
[8] Hurst H 1951 *Trans. Am. Soc. Civil Eng.* **116** 770-84
[9] García L and Carreras B A 2005 *Phys. Plasmas* **12** 092305
[10] Sánchez R Newman D E and Carreras B A 2001 *Nucl. Fusion* **41** 247
[11] Woodard R, Newman D E, Sánchez R and Carreras B A 2007 *Physica A* **373** 215

Journal of Physics: Conference Series **775** (2016) 012009 doi:10.1088/1742-6596/775/1/012009

Uncertainty estimation and a stopping rule in nonlinear gyrokinetic simulations

Michael Oberparleiter, Hans Nordman
Department of Earth and Space Sciences, Chalmers University of Technology, SE-412 96 Gothenburg, Sweden
E-mail: michael.oberparleiter@chalmers.se

Geert Verdoolaege
Research Unit Nuclear Fusion - Department of Applied Physics, Ghent University, Sint-Pietersnieuwstraat 41, B-9000 Ghent, Belgium

Frank Jenko
Department of Physics and Astronomy, University of California, Los Angeles, CA 90095, USA

Abstract. We present a method to estimate the mean and uncertainty of fluctuating quantities, such as spatially averaged density and temperature fluctuations or radial fluxes, from initial value simulations of the Eulerian gyrokinetic code GENE[1, 2]. Since the time series are autocorrelated in time, the data is grouped into batches based on the autocorrelation time and their means form the sample for further statistical treatment, such as calculating the standard error of the mean.Based on this uncertainty estimate we develop a stopping rule for a nonlinear simulation: First, regression tests ensure that it has reached a stationary (quasi-steady) state and data before this point is discarded. Then the previously described estimate is calculated. If the estimated relative error is below a prescribed threshold, the simulation is stopped.

This scheme is applied to several previously performed GENE simulations ranging from simple benchmarks to modelling of JET and ASDEX discharges. It can be demonstrated that a number of simulations could be around 30% shorter if a maximal statistical relative uncertainty of 5% is desired for all monitored quantities.

1. Introduction

Most modern-day investigations of turbulence in tokamak or stellarator plasmas are performed with massively parallel simulation codes which solve the gyrokinetic Vlasov-Poisson system[3]. This can be done by tracking particles in a Lagrangian framework (particle-in-cell, PIC[4, 5]) or by solving the gyrokinetic equations on a fixed grid (Eulerian approach [2, 6–8]. A semi-Lagrangian approach is also successfully used [9].

Although the availability of computing resources has grown significantly over the last years, nonlinear gyrokinetic simulations of turbulence remain an expensive undertaking and an efficient usage of the available core hours is important. In addition, it is preferable to use the additional computational resources to include more physical phenomena in the simulation model and make

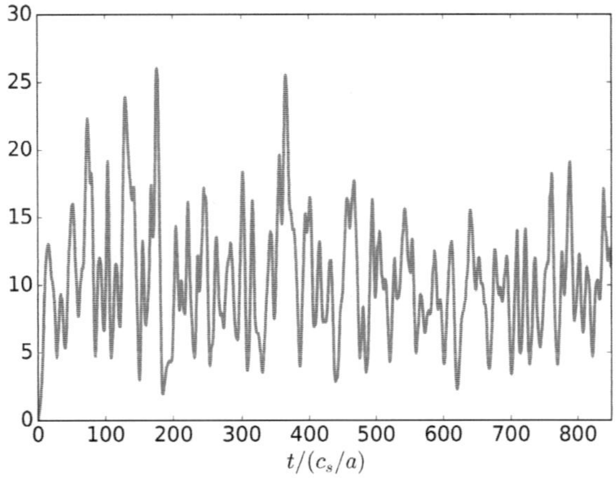

Figure 1: Basic example of a simulation time trace (shown is the flux-surface averaged radial ion heat flux).

it more realistic. Hence, we wish to find a criterion when a simulation can be automatically terminated because it provides results with a sufficient certainty, i.e. with error bars of the size we want.

This immediately raises the question how the (un-)certainty of a simulation result can be estimated. The nonlinear gyrokinetic equation is generally solved numerically as an initial value problem, i.e. simulations provide a time series of the 5D gyrocenter distribution function f and its velocity space moments, which often are physically observable quantities such as density and temperature fluctuations or radial transport fluxes. By taking additional spatial averages they can be reduced to one-dimensional time series which are the most convenient to monitor the state of the simulation.

In Section 2 we discuss one method to find a reliable estimate for the mean and its error of such a time series. Afterwards, in Section 3 we demonstrate how this estimate can be used to establish a stopping rule, i.e. a convergence criterion, for nonlinear simulations.

2. Estimating the mean and uncertainty of fluctuating quantities

In Fig. 1 we show a basic example of a simulation time trace from GENE. Although the figure shows a solid line and the underlying equation systems operate on continuous functions, it is important to remember that the numerical solution and hence our time series are discrete data. In order to apply statistical methods two features are important: First, the time trace contains an initial non-stationary phase ($t \lesssim 50$) which we disregard for the moment and will discuss in Section 3. Secondly, two nearby points in time are not statistically independent, i.e. the data is *autocorrelated.*

We can quantify this by calculating the autocorrelation time t_c, i.e. the time until the normalized discrete autocorrelation function $R(t_i)$ of the time series Q_n has declined to 1/e:

$$R(t_i) = \sum_{j=0}^{N-j} q_i q_{i+j} \quad \text{with } q_i = Q_i - \langle Q \rangle, \tag{1}$$

$$R(t_c) = R(t_0)/\mathrm{e}, \tag{2}$$

where $\langle Q \rangle$ is the average over the entire time trace between the start t_0 and end time t_N. As the samples are not necessarily equidistant in time, we perform averages with the composite

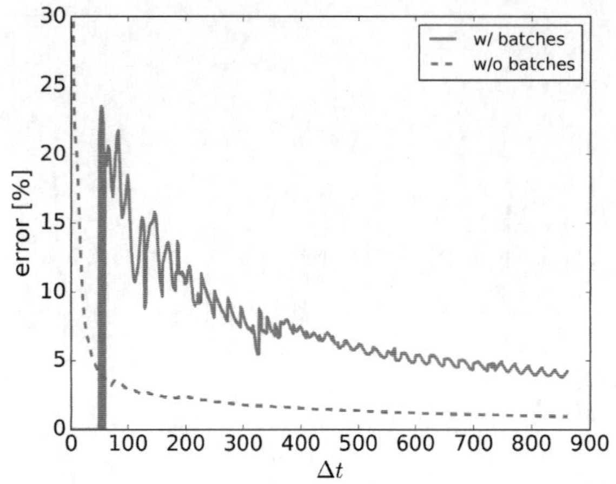

Figure 2: Relative SEM with and without batch mean procedure in dependence of the time interval considered from the time series in Fig. 1.

trapezoidal rule for integration:

$$\langle Q \rangle = \frac{1}{2(t_N - t_0)} \sum_{i=0}^{N-1} (Q_{i+1} + Q_i)(t_{i+1} - t_i). \tag{3}$$

We now use the estimate of the autocorrelation time to construct an uncorrelated set of data: The time interval is separated into N_{win} non-overlapping windows of length $5t_c$ and the data in each window forms a so-called *batch*[10]. This method is also applied, for example, in the field of Markov chain Monte Carlo (MCMC) simulations and has shown a number of good mathematical properties there[11]. The factor 5 ensures that estimation errors of t_c do not lead to remaining autocorrelations. The means $\overline{Q}_n$ of these batches then form the desired set of uncorrelated measurements. The grand mean $\langle Q \rangle$ remains unaffected by this due to its linearity in Q_i and we consider it as the estimate for the average value of the quantity.It is now also straightforward to estimate an uncertainty for said value from the batch means: The standard error of the mean (SEM) from the batch means is defined as:

$$s_{\text{batch}} = \sqrt{\frac{1}{N_{\text{win}}(N_{\text{win}} - 1)} \sum_{n=1}^{N_{\text{win}}} \left(\overline{Q}_n - \langle Q \rangle\right)^2} \tag{4}$$

When we apply this procedure to the data of Fig. 1 and look the relative error $s_{\text{batch}}/\langle Q \rangle$ as a function of the length of the time interval considered (Fig. 2), the SEM shows the expected scaling with $N_{\text{win}}^{-1/2}$, i.e. the error becomes smaller the longer the simulation runs. This sets it apart from, for example, the standard deviation of the data which is sometimes used. As an artefact of the batch process additional small oscillations occur but at an amplitude that is not problematic for relative SEM below 10%. Comparison with a naive calculation of the SEM from the raw data on the other hand demonstrates that ignoring the temporal autocorrelation would lead to a massive underestimate of the uncertainty by a factor of 3-4.

Thus, the relative standard error of the mean provides a good quantity to set a threshold value for stopping the simulation. Typically, a threshold of 5% or 10% is a good choice.

3. Regression testing and the algorithm for the stopping rule

With the uncertainty estimator for autocorrelated time series established the remaining condition that should be ensured is *stationarity*. Generally, it can be violated in two ways for gyrokinetic

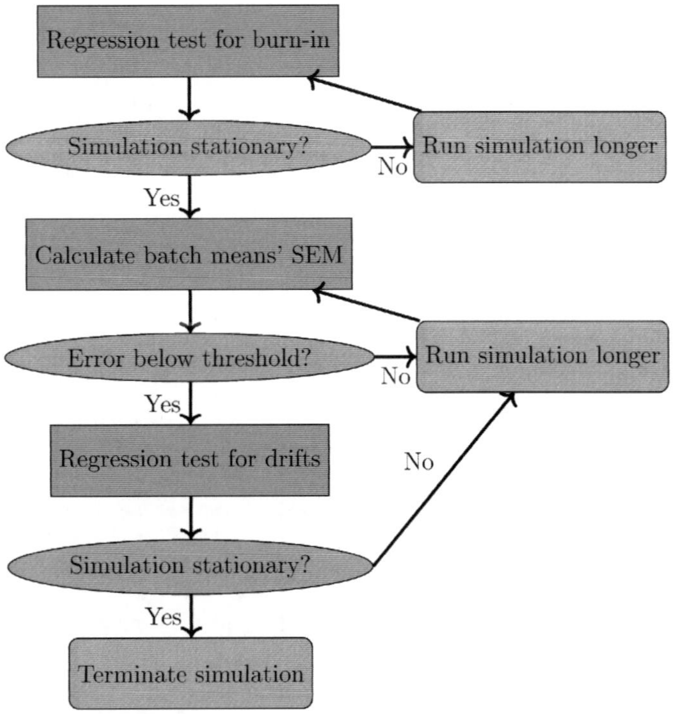

Figure 3: Algorithm for the stopping rule

simulation data: At the start of the simulation the system needs to transition from the initial condition to the quasi-steady state. Typically this occurs as a phase of exponential growth until the nonlinear saturation mechanisms set in. Strictly speaking, it is not necessary for the batch mean procedure to cut off this phase, i.e. allow a "burn-in"[12]. Not doing it, however, significantly increases the required simulation time until the threshold for the SEM is met, as it creates a group of "outliers" in the statistic. The second form of non-stationarity are slow drifts, which can occur due to a number of physical and numerical reasons. In this case, it is preferable to let the code user decide if they consider their simulation run finished.

Both of these issues can be addressed with regression testing. For the burn-in, the last[1] $t_1 - t_0$ time units ($100c_s/a$ in our case) of the simulation are taken and an ordinary least squares linear regression is performed, resulting in the linear equation

$$Q(t) = \alpha(t - t_0) + \beta. \tag{5}$$

If $|\alpha/(t_1 - t_0)| > 0.1\langle Q\rangle$, i.e. the regression line's slope is larger than 0.1 of the data mean in the regression interval or, in case the mean is smaller than the data's standard deviation σ, $|\alpha/(t_1 - t_0)| > 1.5\sigma$, the test fails. As soon as the test succeeds for the first time, the start of the 100 time unit interval is set as the starting point for the procedure of Section 2.

When the threshold for the uncertainty estimate is reached, the regression test is applied again. We, however, perform the regression over the entire time series (without the burn-in) and allow for the slope to be as large as 0.2 of the mean. If this test passes, the simulation can be stopped. The overview over the entire algorithm is shown in Fig. 3.

[1] This is from the perspective of a running simulation. A posteriori, this analysis can be done by taking an initial interval of data and extending it every time the test fails.

Journal of Physics: Conference Series **775** (2016) 012009 doi:10.1088/1742-6596/775/1/012009

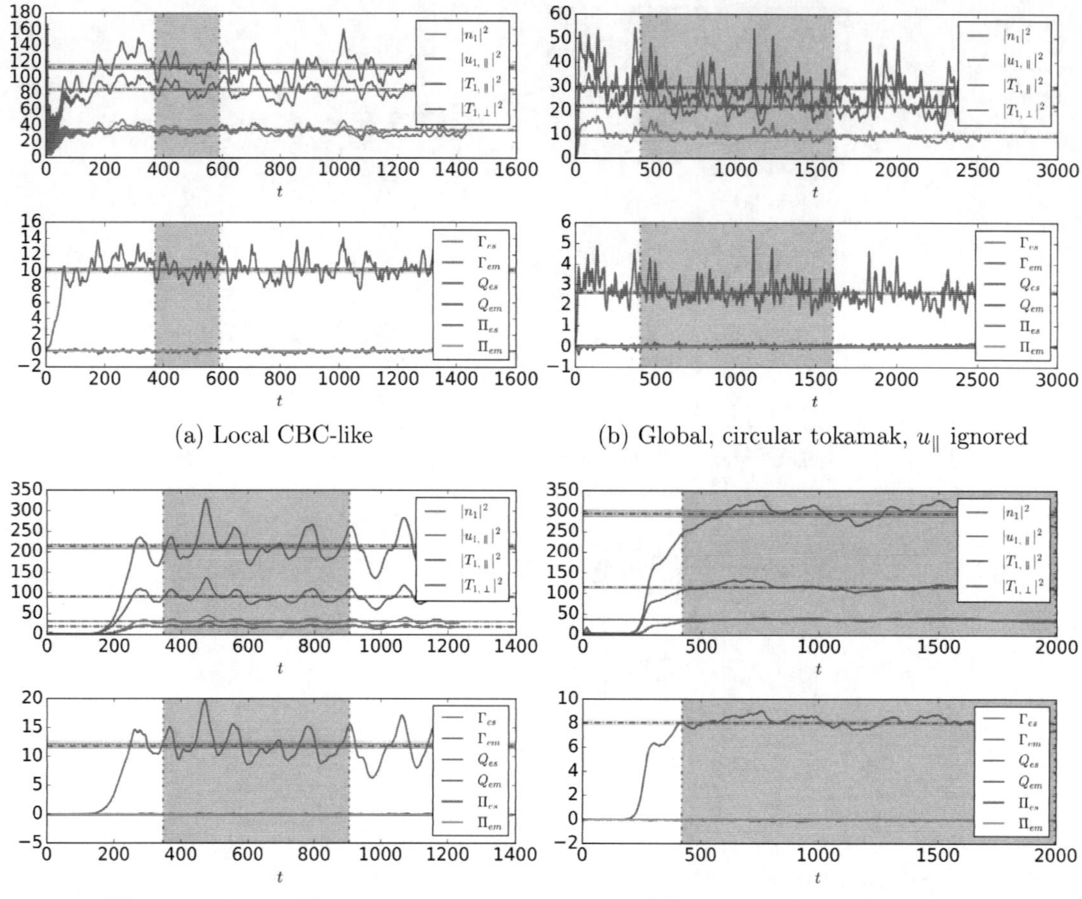

(a) Local CBC-like

(b) Global, circular tokamak, $u_{\parallel}$ ignored

(c) Full-surface NCSX geometry

(d) Full surface W7X geometry

Figure 4: ITG turbulence simulations with adiabatic electrons

4. Applications to simulation data

The straightforward way to test this algorithm is to a posteriori apply it to a diverse set of already finished simulation runs. In the following we present time traces for the spatially averaged fluctuations of density $|n_1|^2$, parallel flow $|u_{1\parallel}|^2$, temperature $|T_{1\parallel}|^2$ and $|T_{1\perp}|^2$, as well as the flux surface averaged flows of particles Γ, energy Q and momentum Π. If a quantity is numerically zero, it is ignored in the analysis. The shaded time intervals show the result of the stopping rule algorithm with an uncertainty threshold of 5% for all quantities. The dashed horizontal lines mark the average of a quantity based on the full time series excluding the burn-in, while the dotted lines show the average based on the stopping rule algorithm.

Fig. 4a shows the most basic type of simulated tokamak turbulence: A local (flux-tube) simulation of a circular tokamak with Cyclone Base Case-like[13] parameters with electrons considered as adiabatic. As can be seen, the stopping rule would terminate the simulation after a surprisingly short time (40% of the total time) with a maximal uncertainty of 3.6% and the mean values agreeing well with their reference from the full time series.

In Fig. 4b the simulation is global, i.e. it takes the radial profiles of density and temperature into account. Otherwise it is similar to the previous case. For the analysis we suppress $u_{\parallel}$, as it has a long time drift that is deemed irrelevant. The error bar reaches its desired value of 4.9% after 65% of the full time series. Since the cost for a global simulation is about an order

Journal of Physics: Conference Series **775** (2016) 012009 doi:10.1088/1742-6596/775/1/012009

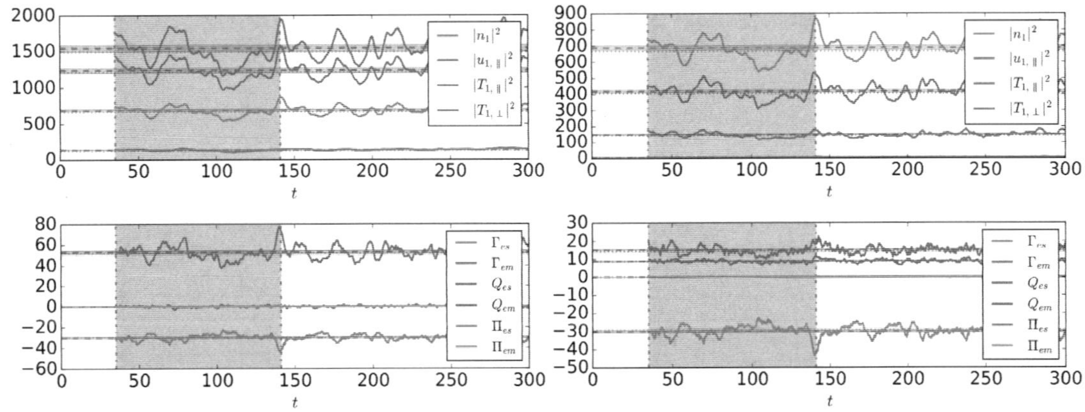

(a) Ions (b) Electrons

Figure 5: Local ITG/TEM turbulence simulation with electromagnetic effects of a JET discharge

of magnitude higher than for a flux-tube, this would imply a greater saving than for the case of Fig. 4a.

Fig. 4c and 4d show time series of full-flux-surface simulations of stellarator geometries. Already from visual inspection it becomes clear that their autocorrelation time is notably longer than previously. For the case of Fig. 4c still an uncertainty of 4.8% is reached after 75% of the simulation time. In the Wendelstein-7X case the final drift regression test fails and the batch means procedure results in an uncertainty of 2.5% for the full time trace.

In all cases presented so far, the regression testing to find the burn-in phase performed completely satisfactory. In the case of a simulation continued from a checkpoint as in Fig. 5, it can, however, cause too much data to be neglected. Hence, we do not apply it in this simulation of a JET discharge. The remaining part of the stopping rule performs well for this physically quite comprehensive modelling and a relative uncertainty of 5.0% is reached after 45% of the used simulation time. It should be noted at this point that for all time series presented so far the average values of the quantities based on the stopping rule agreed with their counterparts based on the full time trace within their uncertainty.

Lastly, we show a simulation of an ASDEX-Upgrade discharge (Fig. 6) where the algorithm does not converge and also does not give a reasonable error estimate. Due to a source of external $\vec{E} \times \vec{B}$ shear switched on at $t = 200$ the time trace is not stationary until fairly late and the remaining data is insufficient for the batch means procedure. In addition, the stationarity of the entire series is not easily to establish. Thus, we deem this one of the scenarios where manual control and evaluation by the code user is the better option.

5. Conclusions

In this article an algorithm was developed to automatically terminate a nonlinear gyrokinetic simulation when it gives results with a predetermined uncertainty, i.e. a previously chosen error bar. For this, the standard error of the mean calculated from (uncorrelated) batch means provides a statistically sound estimate of the uncertainty. It can be combined with linear regression tests for better convergence properties and more reliable average values derived from the simulation. It could be shown in an a posteriori analysis of a number of simulation runs from GENE that significant savings in simulation time are possible with this algorithm. There remain, however, a number of scenarios which require further fine tuning of the method and its parameters.

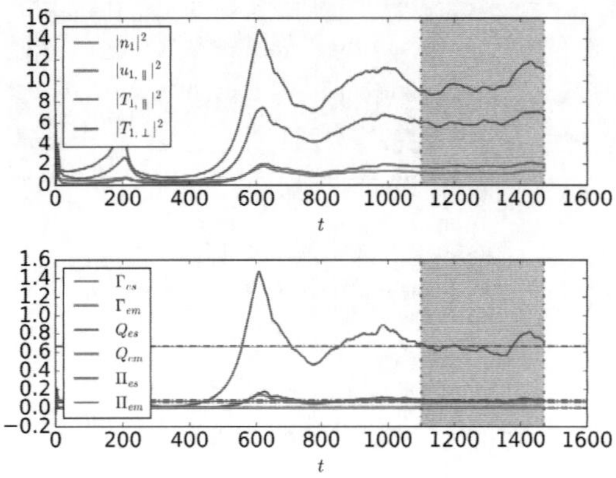

(a) Thermal ions

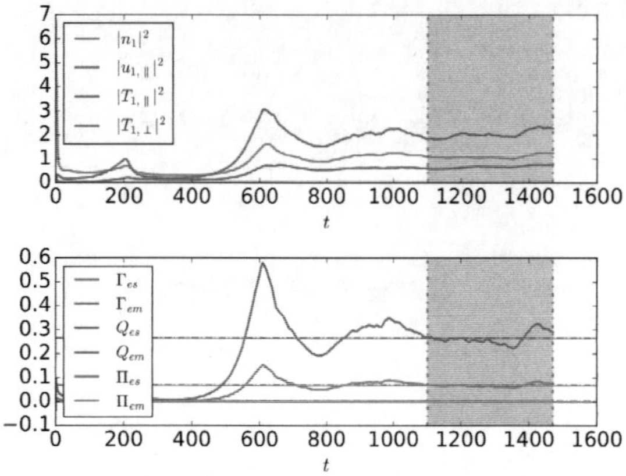

(b) Electrons

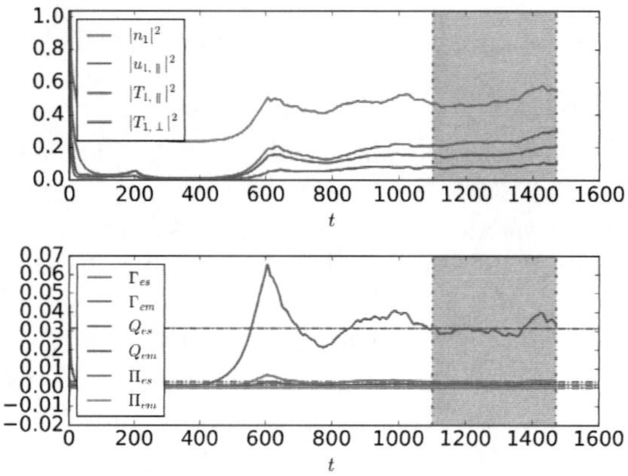

(c) Fast ions

Figure 6: Local ITG/TEM turbulence simulation with a fast particle species and external $\vec{E} \times \vec{B}$ shear for an AUG discharge

6. Acknowledgements

The authors would like to thank A. Bustos, T.Görler, D.Told, M.J.Pueschl, H.Doerk and D.Tegnered for providing significant parts of the simulation data analysed in this work.

References

[1] Jenko F, Dorland W, Kotschenreuther M and Rogers B N 2000 *Phys. Plasmas* **7** 1904–1910

[2] Görler T, Lapillonne X, Brunner S, Dannert T, Jenko F, Merz F and Told D 2011 *J. Comput. Phys.* **230** 7053 – 7071 URL `http://genecode.org`

[3] Brizard A J and Hahm T S 2007 *Reviews of Modern Physics* **79**(2) 421–468

[4] Kornilov V, Kleiber R, Hatzky R, Villard L and Jost G 2004 *Phys Plasmas* **11** 3196

[5] Jolliet S, Bottino A, Angelino P, Hatzky R, Tran T, Mcmillan B, Sauter O, Appert K, Idomura Y and Villard L 2007 *Comput. Phys. Commun.* **177** 409 – 425 ISSN 0010-4655

[6] Jenko F, Dorland W and Hammett G W 2001 *Phys. Plasmas* **8** 4096–4104

[7] Peeters A, Camenen Y, Casson F, Hornsby W, Snodin A, Strintzi D and Szepesi G 2009 *Comput. Phys. Commun.* **180** 2650–2672

[8] Idomura Y, Urano H, Aiba N and Tokuda S 2009 *Nucl. Fusion* **49** 065029

[9] Grandgirard V, Brunetti M, Bertrand P, Besse N, Garbet X, Ghendrih P, Manfredi G, Sarazin Y, Sauter O, Sonnendrücker E, Vaclavik J and Villard L 2006 *Journal of Computational Physics* **217** 395 – 423 ISSN 0021-9991

[10] Law A M and Kelton W D 1991 *Simulation Modeling and Analysis* 2nd ed (McGraw-Hill)

[11] Flegal J M and Jones G L 2010 *Ann. Statist.* **38** 1034–1070

[12] Jones G L, Haran M, Caffo B S and Neath R 2006 *J. Amer. Statist. Assoc.* **101** 1537–1547

[13] Dimits A M, Bateman G, Beer M A, Cohen B I, Dorland W, Hammett G W, Kim C, Kinsey J E, Kotschenreuther M, Kritz A H, Lao L L, Mandrekas J, Nevins W M, Parker S E, Redd A J, Shumaker D E, Sydora R and Weiland J 2000 *Phys. Plasmas* **7** 969–983

Towards the optimization of a gyrokinetic Particle-In-Cell (PIC) code on large-scale hybrid architectures

N Ohana[1], A Jocksch[2], E Lanti[1], T M Tran[1], S Brunner[1], C Gheller[2], F Hariri[1], L Villard[1]

[1] Swiss Plasma Center, Ecole Polytechnique Fédérale de Lausanne, Switzerland
[2] Swiss National Supercomputing Centre, Lugano, Switzerland

E-mail: `noe.ohana@epfl.ch`

Abstract. With the aim of enabling state-of-the-art gyrokinetic PIC codes to benefit from the performance of recent multithreaded devices, we developed an application from a platform called the "PIC-engine" [1, 2, 3] embedding simplified basic features of the PIC method. The application solves the gyrokinetic equations in a sheared plasma slab using B-spline finite elements up to fourth order to represent the self-consistent electrostatic field. Preliminary studies of the so-called Particle-In-Fourier (PIF) approach, which uses Fourier modes as basis functions in the periodic dimensions of the system instead of the real-space grid, show that this method can be faster than PIC for simulations with a small number of Fourier modes. Similarly to the PIC-engine, multiple levels of parallelism have been implemented using MPI+OpenMP [2] and MPI+OpenACC [1], the latter exploiting the computational power of GPUs without requiring complete code rewriting. It is shown that sorting particles [3] can lead to performance improvement by increasing data locality and vectorizing grid memory access. Weak scalability tests have been successfully run on the GPU-equipped Cray XC30 Piz Daint (at CSCS) up to 4,096 nodes. The reduced time-to-solution will enable more realistic and thus more computationally intensive simulations of turbulent transport in magnetic fusion devices.

1. Introduction

In order to use nuclear fusion as a source of energy, magnetic confinement of plasmas needs to be improved. Experimental devices are facing difficulties associated with so-called *anomalous transport*, which is attributed to microturbulence. Because of the physical complexity and the chaotic behavior of the system, analytical studies are limited and numerical schemes are necessary to better understand instabilities occurring in it.

During the last decades, much effort has been invested into the development of first principle models, among which we find gyrokinetic Particle-In-Cell (PIC) codes. They apply a Monte-Carlo particle method to gyrokinetic theory, which takes advantage of the difference in time scales between the fast cyclotron motion of charged particles in a magnetic field and the slow instabilities. The phase space is thus reduced to five dimensions, and simulations can use larger time steps. The PIC method then consists of discretizing the distribution functions of each species with numerical markers, and representing the fields on a real space grid, e.g. with finite elements.

Despite the use of many numerical methods to reduce the noise inherent to particle methods (e.g. δf scheme [4], noise control [5], ...), we are still too limited by numerical resources to achieve the accuracy required to explain experimental observations. For instance, simulations of a medium-sized tokamak taking into account certain effects related to electron dynamics together with ion dynamics require a grid of at least $360 \times 1024 \times 512$ cells, and 2 billion markers of each species [6]. Such resolution for large-sized devices reaches the limits of today's computational resources.

We thus have to turn towards High Performance Computing (HPC), a field in constant evolution. Among the arising multi-threaded architectures, a trend in scientific computing is to use General-Purpose computation on Graphic Processing Units (GPGPU). GPUs turn out to be particularly suitable for PIC codes [7], thanks to the high number of particles that can be treated simultaneously by the device. The OpenACC directive-based interface is particularly interesting for portability purposes as one can run the same code on different architectures.

In this paper, we present the so-called "GK-engine", a code that embeds all the main features of a gyrokinetic PIC code in a modular and portable structure. It can be used as a testbed for various numerical methods or data structures on different platforms, without requiring the refactoring of all the physics modules around the core. As an example, a variant of the classical representation of fields on a real-space grid was easily implemented, the so-called PIF approach. It stores the Fourier components of fields in the toroidal and poloidal dimensions, in order to get compact field data and intrinsic filtering of numerical noise.

The paper is organized as follows: section 2 describes the simplified GK-engine model and its implementation; section 3 shows the physical validation of the code by application to Ion Temperature Gradient (ITG) instabilities; section 4 checks the numerical convergence of the simulations; section 5 presents the performance of the different architectures and implementations, and section 6 looks at the parallel scalability of the code.

2. The GK-engine model

For simplicity, we consider a slab-like system with Cartesian coordinates. x will represent the radial direction, y the poloidal one, and z the toroidal one. The magnetic field can thus be written as $\vec{B}(x,y,z) = B_y(x)\hat{e}_y + B_z\hat{e}_z$. We have arbitrarily chosen to include the radial dependency in the poloidal component exclusively. The safety factor $q(x) = \frac{L_y}{L_z}\frac{B_z}{B_y}$ is a measure of the magnetic field tilt angle, where L_i stands for the length of dimension i.

Gyrokinetic theory consists of following the guiding center dynamics in the 5D phase space $(\vec{R}, v_{/\!/}, \mu)$, where $\vec{R}$ stands for the position, $v_{/\!/}$ the projection of velocity on the magnetic field direction, and $\mu = \frac{mv_\perp^2}{2B}$ the magnetic moment with m the particle mass, $v_\perp$ the perpendicular velocity, and B the magnetic field strength. Within our geometry, the equations of motion are:

$$\frac{d\vec{R}}{dt} = v_{/\!/}\hat{b} + \frac{\mu}{eB^\star}\hat{b} \times \vec{\nabla} B + \frac{1}{B^\star}\hat{b} \times \vec{\nabla}\langle\phi\rangle \tag{1}$$

$$\frac{dv_{/\!/}}{dt} = -\frac{e}{m}\hat{b} \cdot \vec{\nabla}\langle\phi\rangle \tag{2}$$

$$\frac{d\mu}{dt} = 0 \tag{3}$$

with $B^\star = B\Big(1 + \frac{mB_y' B_z}{eB^3}v_{/\!/}\Big)$, $\hat{b}$ the unitary magnetic field vector, e the particle charge, ϕ the electrostatic potential, $\langle\cdot\rangle$ standing for gyroaveraging and the prime $'$ for $\partial/\partial x$.

The electrostatic field is solved using the quasineutrality equation with the long wavelength approximation and adiabatic electrons:

$$\frac{en_0}{T_e}(\phi - \bar{\phi}) - \vec{\nabla}_\perp \cdot \left(\frac{n_0}{B\Omega_i}\vec{\nabla}_\perp \phi\right) = \int \frac{B^\star}{m} f(\vec{x} + \vec{\rho}_L(\mu,\alpha), v_{/\!/}; \mu)\, d\mu\, d\alpha\, dv_{/\!/} - n_0 \tag{4}$$

Journal of Physics: Conference Series **775** (2016) 012010 doi:10.1088/1742-6596/775/1/012010

where T_e is the electron temperature, $\bar{\cdot}$ stands for the flux-surface averaging (i.e. in toroidal and poloidal directions), Ω_i is the ion cyclotron frequency, $\vec{\rho}_L$ is the Larmor radius, and f is the ion distribution function. The right-hand-side term is the gyro-averaged density.

f is decomposed into an equilibrium background part f_0 that does not depend on time, and a perturbed part δf that we discretize with numerical markers:

$$\delta f(\vec{R}, v_{/\!/}, t; \mu) = \sum_{p=1}^{N} \frac{1}{2\pi B^\star} w_p(t)\delta\big(\vec{R} - \vec{R}_p(t)\big)\delta\big(v_{/\!/} - v_{/\!/p}(t)\big)\delta\big(\mu - \mu_p\big) \tag{5}$$

where w_p are the marker weights. Markers are initialized to populate uniformly the phase space (velocity space is delimited by $\kappa_v v_{th}$, with κ_v typically chosen around 5), and their weights are initialized with:

$$w_p(t=0) = \underbrace{\frac{L_x L_y L_z v_{\perp p}[\pi \kappa_v v_{th}(x_p)]^2 B^\star(x_p, v_{/\!/p})}{N B(x_p)}}_{\Omega_p} \delta f_p(t=0) \tag{6}$$

where Ω_p is the phase space volume occupied by marker p.

The time evolution of the weights is given by the Vlasov equation:

$$\frac{dw_p}{dt} = \Omega_p \frac{d\delta f_p}{dt} = -\Omega_p \left(\frac{d\vec{R}}{dt} \cdot \vec{\nabla} f_0 + \frac{dv_{/\!/}}{dt} \frac{\partial f_0}{\partial v_{/\!/}} \right)_p \tag{7}$$

We use a Runge-Kutta scheme of order 4 for this evolution. Each Runge-Kutta substep is composed of the successive stages shown in Figure 1.

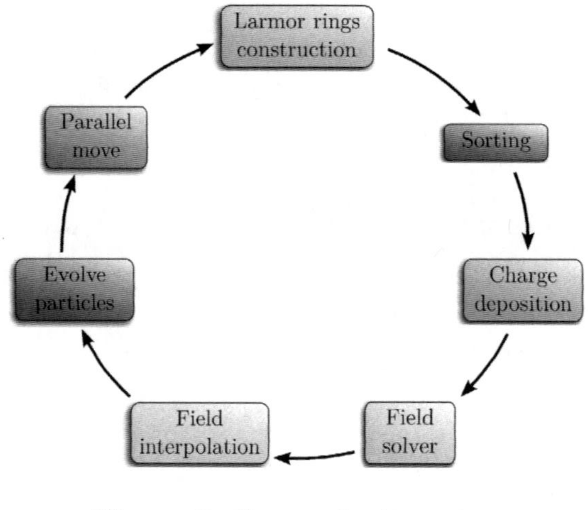

Figure 1. Stages of a time step.

The first stage is the construction of Larmor rings from guiding centers, using an adaptive number of points for each guiding center proportional to its Larmor radius ρ_L.

Then, one needs to build the gyro-averaged density in order to solve equation (4). This is done by depositing the charge on a finite element basis, in our case B-splines of order p [8]:

$$\rho(x,y,z) = \sum_{i=1}^{N_x+p} \sum_{j=1}^{N_y+p} \sum_{k=1}^{N_z+p} \hat{\rho}_{ijk} \Lambda_i(x) \Lambda_j(y) \Lambda_k(z) \tag{8}$$

where N_i is the number of intervals along dimension i and $\{\Lambda_i\}$ the B-splines functions. Thanks to the locality of finite elements, one only needs to sum over a restricted number of splines for a given position in space. For numerical performance purposes, it can be useful to sort the markers according to the grid cell in which they reside. Indeed, it will increase data locality and allow to loop first on the grid cells and then on the particles in the cell. Otherwise, if the particles are not sorted, one would simply loop on particles and randomly access the grid data.

It is then convenient to perform Discrete Fourier Transforms (DFTs) of the field coefficients $\{\hat{\rho}_{ijk}\}$ along y and z directions, allowing us to filter out numerical noise [9]. Knowing that the instability wavelengths are much longer in the parallel direction than the perpendicular one [10], one can set to 0 the modes with a poloidal wavenumber m such that $|m + nq(x)| > \Delta m$, where n is the toroidal wavenumber. Δm is a parameter typically set to 5.

Alternatively, the PIF method directly builds the field Fourier transform along y and z:

$$\rho(x, y, z) = \sum_{i=1}^{N_x+p} \sum_{n=n_{min}}^{n_{max}} \sum_{m=-nq_i-\Delta m}^{-nq_i+\Delta m} \hat{\rho}_{imn} \Lambda_i(x) e^{2\pi i(my/L_y+nz/L_z)} \tag{9}$$

where q_i is the safety factor evaluated at the middle of the support of spline Λ_i.

One can then solve the quasineutrality equation for the electrostatic potential, and evaluate this field at each marker position to make them evolve according to equations 1, 2 and 7.

In the PIC implementation, the parallelization scheme of the GK-engine allows for domain decomposition along the z direction, i.e. splitting the grid data into subdomains distributed among MPI tasks. The markers of each processor have to be in the right subdomain, therefore one needs an MPI communication stage (called parallel move) to send markers leaving a subdomain to the new one. An extra dimension of parallelism consists of cloning the subdomains on several MPI tasks in order to reduce the number of particles per node.

3. ITG simulations

As a validation step, we choose to simulate ITG instabilities arising in a tokamak of major radius $R_0 = 77$ cm and minor radius $a = 21$ cm. Our slab model uses $L_x = a$, $L_y = \pi a$ and $L_z = 2\pi R_0$. The magnetic field on axis is $B_0 = 1$ T, and the safety factor is given by:

$$q(s) = 1.25 + 3s^2 \tag{10}$$

where $s = x/a$ is the normalized radial coordinate.

The temperature and density profiles are given by:

$$T_i(s) = T_{i0} e^{-\kappa_{T_i} \Delta_{T_i} \tanh\left(\frac{s-0.5}{\Delta_{T_i}}\right)} \tag{11}$$

$$T_e(s) = T_{e0} e^{-\kappa_{T_e} \Delta_{T_e} \tanh\left(\frac{s-0.5}{\Delta_{T_e}}\right)} \tag{12}$$

$$n_0(s) = n_{00} e^{-\kappa_n \Delta_n \tanh\left(\frac{s-0.5}{\Delta_n}\right)} \tag{13}$$

with the parameters $T_{i0} = T_{e0} = 1$ keV, $\kappa_{T_i} = 2$, $\Delta_{T_i} = 0.2$, $\kappa_{T_e} = 1$, $\Delta_{T_e} = 0.1$, $\kappa_n = 0.5$, and $\Delta_n = 0.3$.

Local linear analytical studies can show that a condition for instability is $\eta(s) := \frac{d\ln(T_i(s))}{d\ln(n(s))} \gtrsim 2$ [15]. A second condition is that $k_{/\!/}\rho_s \ll 1$, where $\rho_s = \frac{\sqrt{mT_e}}{eB}$ is the ion sound Larmor radius.

We first run linear simulations using the PIF method with 5M particles and 128 radial cubic splines in order to find the most unstable mode of the system (see Figure 2).

We find that the growth rate γ is positive only for $n \lesssim 13$, and that the most unstable modes satisfy $m = -nq(s = 0.5)$, i.e. $k_{/\!/}(s = 0.5) = 0$.

Journal of Physics: Conference Series **775** (2016) 012010 doi:10.1088/1742-6596/775/1/012010

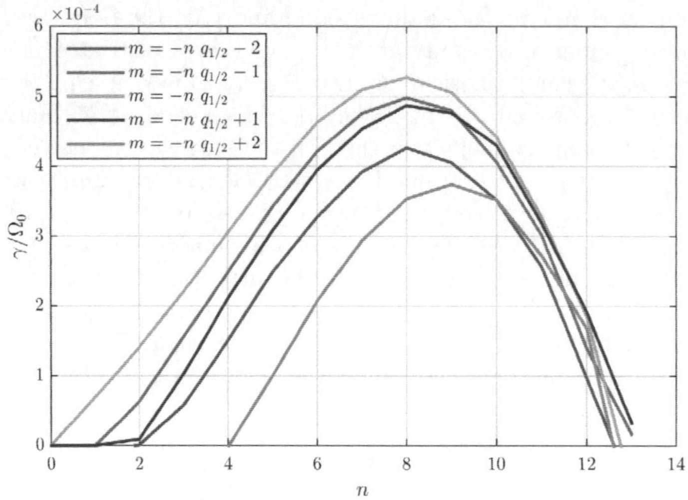

Figure 2. Linear growth rate of quasi field-aligned Fourier modes. Simulations with PIF method, 5M particles, and 128 radial cubic splines.

Let us now initialize the overall most unstable mode $(m, n) = (-16, 8)$, and simulate until nonlinear phase begins. We use the PIC method with $\Delta t = 10/\Omega_0$, 200M particles and $128 \times 512 \times 128$ cubic splines (see Figure 3).

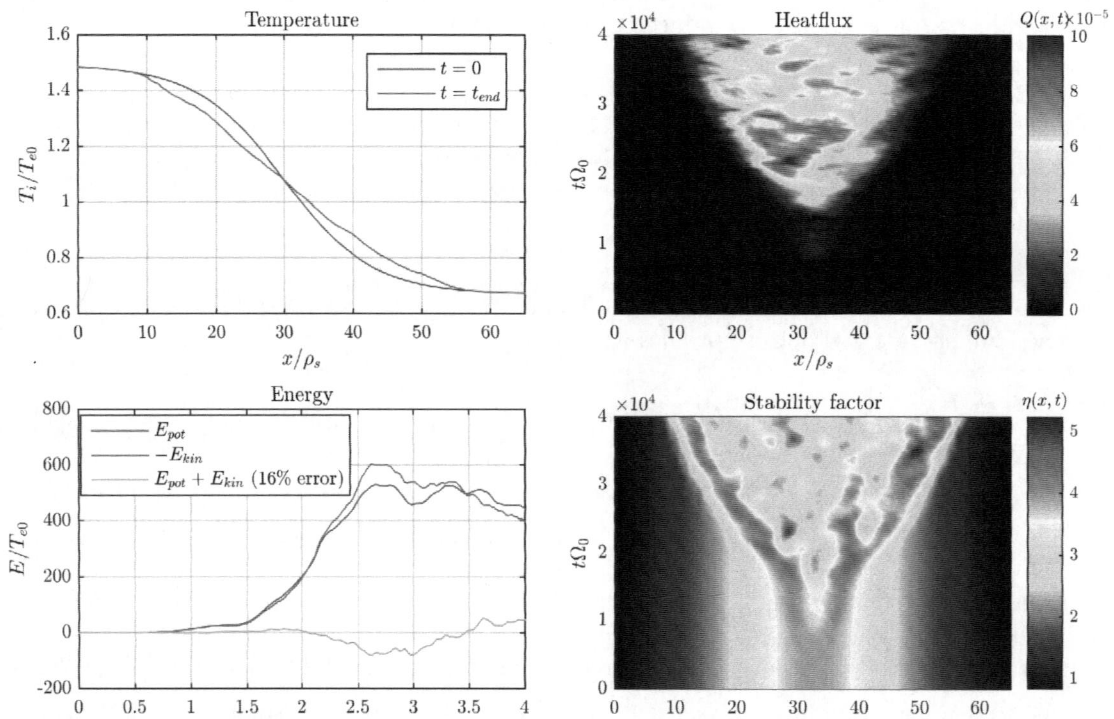

Figure 3. Diagnostics of a nonlinear simulation with PIC method, $\Delta t = 10/\Omega_0$, 200M particles, $128 \times 512 \times 128$ cubic splines. Fourier filter only keeps $0 \leq n \leq 25$ and $|m + nq| \leq 5$.

We see that starting at $t \approx 10^4/\Omega_0$, an outward radial heat flux flattens the temperature profile. The kinetic energy is converted into potential energy, and the balance is satisfied.

4. Convergences studies

Having verified that our simulations are physically meaningful, we want to investigate the numerical convergence of the results.

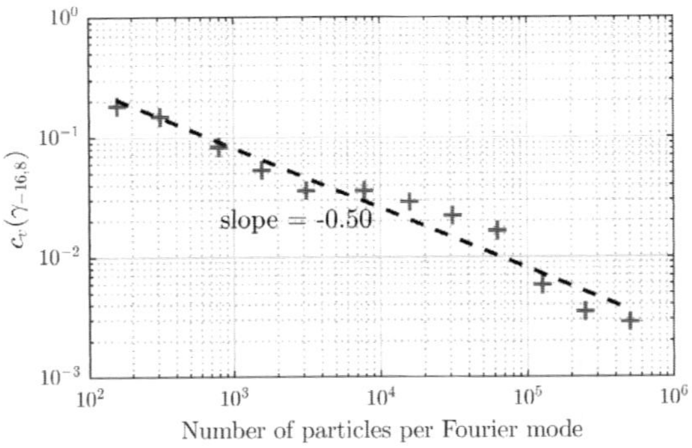

Figure 4. Coefficient of variation in time of the linear growth rate as a function of the number of particles. Simulations with PIF, $\Delta t = 200/\Omega_0$, and 128 radial cubic splines.

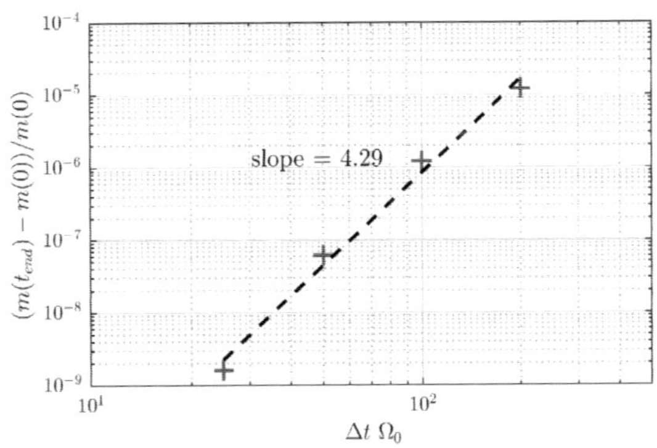

Figure 5. Mass non-conservation at $t_{end} = 4 \cdot 10^4\ \Omega_0$ of a nonlinear simulation as a function of the time step. Simulations with PIF, 100M particles, and 128 radial quadratic splines.

In Figure 4, we look at the linear growth rate of a single mode as a function of the number of particles. As we find that the instantaneous growth rate temporally varies around an average value, we plot the coefficient of variation of this oscillation, i.e. the ratio of the standard deviation to the average. This quantity is found to be inversely proportional to the square root of the number of particles, reflecting a statistical noise.

Figure 5 reports the relative variation of the total mass ($\iint f d^3x\ d^3v$) at the end of the nonlinear simulation. It converges with Δt^4, consistent with the time integrator we used (fourth order Runge-Kutta).

5. Numerical optimization

We now run a benchmark of the different implementations of the code, on CPU and on GPU, for a simulation of a unique Fourier mode (see Figure 6). On CPU, we can either use 8 MPI tasks, or a single task with 8 OpenMP threads. Unsorted and sorted versions are compared; loops on particles are better vectorized in the latter version because sorting increases data locality and avoids random memory access. Moreover, for the multithreaded implementations, sorting enables the multithreaded outer loop to be on the grid cells.

One can see that the most important steps are particle-to-grid and grid-to-particle operations. On CPU implementations, the sorting is more expensive than the speed-up it brings, but on GPU it gives an overall gain of a factor 2 because it avoids conflicts between multiples threads accessing the same grid point in memory (race condition). The sorting routine could still be optimized using another algorithm [3], or taking advantage of the fact that Larmor rings lie on the same poloidal plane.

For comparison, we also show the best implementation of the PIF method, namely on GPU with sorting. It is twice faster than the best PIC implementation.

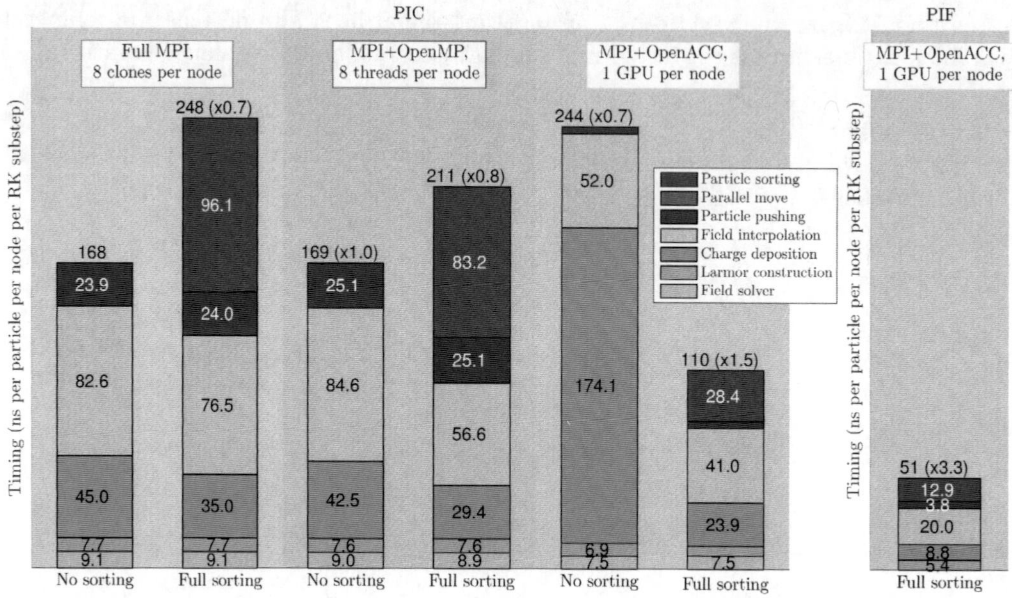

Figure 6. Timings comparison of full MPI, OpenMP, and OpenACC implementations, with or without sorting, on 1 node of Piz Daint. Parameters are 2M particles and $64 \times 64 \times 32$ cubic splines for PIC, and 1 mode for PIF.

Then, we run the same benchmarking on a bigger case with 26×11 Fourier modes (Figure 7).

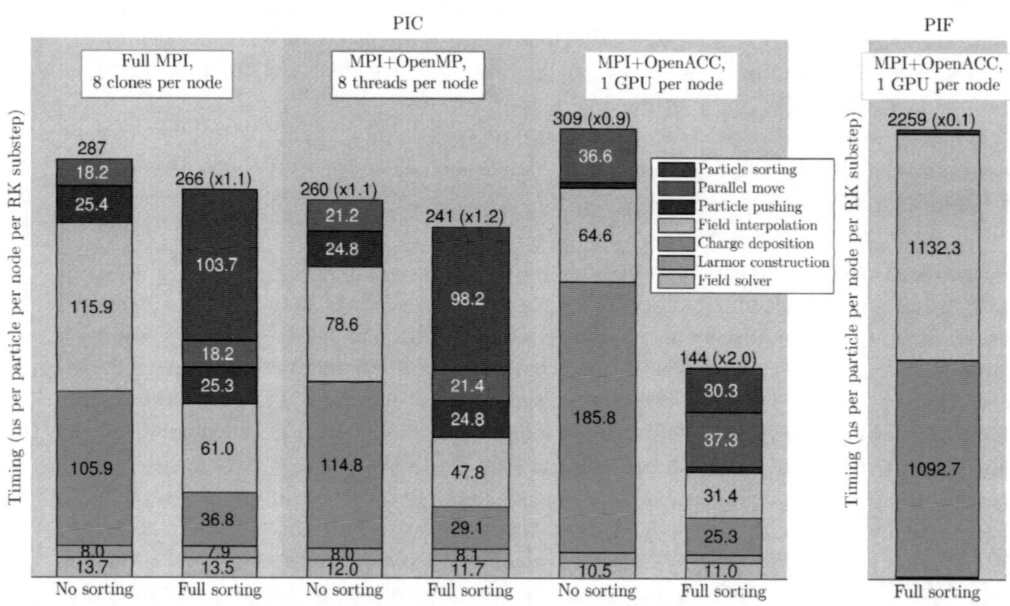

Figure 7. Same comparison as Figure 6 with 32 subdomains (32 nodes). Parameters are 32 subdomains, 100M particles, $128 \times 512 \times 128$ cubic splines for PIC, and 32 clones, 26×11 modes for PIF (shown on a different scale).

This time, sorting particles is always worth it, even on CPU implementations. We also see the parallel move timing, slower on the GPU version because it requires a lot of host-device communications. The overall best timing is achieved by the GPU implementation with sorting.

The PIF method is now one order of magnitude slower than PIC, because all the Fourier modes need to be evaluated at each marker position. Although optimization is still under development, it seems that this method will only be advantageous for a restricted number of Fourier modes.

6. Parallel scalability

Lastly, we run a weak scaling to get an idea of the behavior of the timings on a large scale cluster. The length of the domain and the time step are fixed, but the toroidal resolution and the number of particles are increased proportionally to the number of nodes.

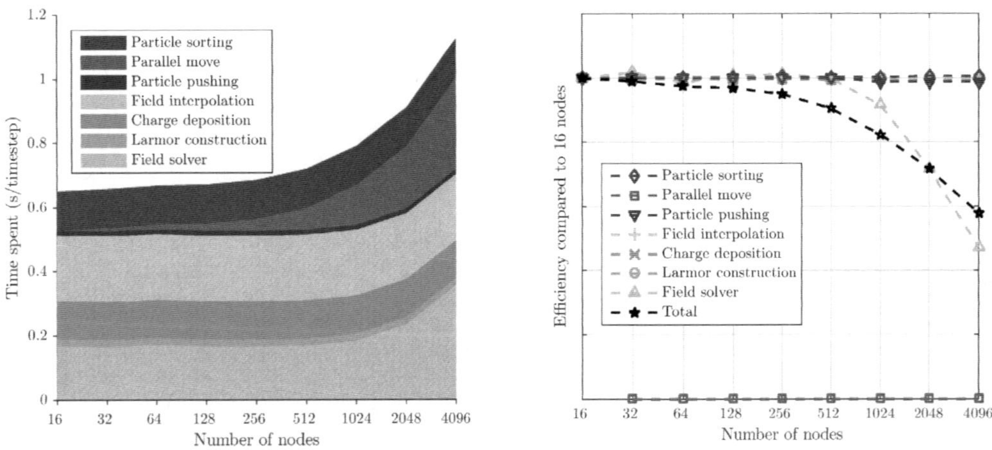

Figure 8. Weak scaling of sorted OpenACC version with 16 clones and 1 to 256 subdomains. Simulation parameters are 10^6 guiding centers per node, $128 \times 512 \times (4$ grid cells per subdomain), and cubic interpolation. Simulations were run on Piz Daint.

The result, shown on Figure 8, clearly demonstrates that most of the routines scale perfectly well (over 98% efficiency).

The exceptions are the field solver and the parallel move. Indeed, when increasing the number of subdomains for a fixed-size domain, the number of escaping particles increases as well, resulting in complete non-scalability. For a large number of subdomains, a large fraction of the particles needs to be sent to other subdomains. We are currently considering adding another direction of domain decomposition to counterbalance this problem.

As for the field solver, the non-scalability comes from the parallel data transposes required to perform Discrete Fourier Transforms (DFTs). In the future, we will investigate an idea to compute partial local DFTs and communicate compacted fields instead of the full real grid.

7. Conclusions

We have demonstrated the proper functioning of the GK-engine both physically, through linear and nonlinear ITG simulations, and numerically through convergence studies. We then compared performance of different implementations (MPI cloning versus OpenMP multithreading, sorting or not) and different architectures (CPU versus GPU) in two reference cases (small-sized on a single node and medium-sized on several nodes). We found that on CPU, sorting can be more costly than the gain it brings, in particular for a small grid, but that it is always profitable on GPU because it avoids race conditions.

We also found that the PIF performance depends greatly on the number of Fourier modes we want to represent. Even though it is not yet fully optimized, we know that its application will be limited to systems where the number of Fourier modes is much smaller than the real-space grid size. It could for instance be interesting for toroidal Alfvén eigenmodes simulations, as one typically needs only one toroidal mode and five or seven poloidal modes, eventually adding the $n = 0$ modes to study the coupling to zonal flows and geodesic acoustic modes.

The knowledge we acquired with the GK-engine will help in the refactoring of the larger gyrokinetic PIC code ORB5 [11, 12, 13].

Acknowledgements

This work is partly funded by the Swiss National Science Foundation and the Platform for Advanced Scientific Computing. It has been carried out within the framework of the EUROfusion Consortium and has received funding from the Euratom research and training programme 2014-2018 under grant agreement No 633053. The views and opinions expressed herein do not necessarily reflect those of the European Commission.

References

[1] Hariri F, Tran T M, Jocksch A, Lanti E, Progsch J, Messmer P *et al.* 2016 A portable platform for accelerated PIC codes and its application to GPUs using OpenACC *Comput. Phys. Commun.* (in press, available online on May 21 2016)

[2] Lanti E *et al.* 2016 Towards porting a gyrokinetic PIC code on many-core platforms *Comput. Phys. Commun.* (submitted)

[3] Jocksch A, Hariri F, Tran T M, Brunner S, Gheller C and Villard L 2015 A bucket sort algorithm for the particle-in-cell method on manycore architectures *International Conference on Parallel Processing and Applied Mathematics* 43-52

[4] Brunner S, Valeo E and Krommes J A 1999 Collisional delta-f scheme with evolving background for transport time scale simulations *Phys. Plasmas* **6**(12) 4504

[5] McMillan B F, Jolliet S, Tran T M, Villard L, Bottino A and Angelino P 2008 Long global gyrokinetic simulations: Source terms and particle noise control *Phys. Plasmas* **15**(5) 052308

[6] Dominski J 2016 Development and application of Eulerian and Particle-In-Cell gyrokinetic codes for studying the effect of non-adiabatic passing electron dynamics on microturbulence *PhD thesis* EPFL

[7] Burau H, Widera R, Honig W, Juckeland G, Debus A, Kluge T *et al.* 2010 PIConGPU: A fully relativistic particle-in-cell code for a GPU cluster *IEEE Transactions on Plasma Science* **38**(10) 2831-9

[8] De Boor C 1978 A practical guide to splines *New York: Springer-Verlag* **27** 325

[9] McMillan B F, Jolliet S, Bottino A, Angelino P, Tran T M and Villard L 2010 Rapid Fourier space solution of linear partial integro-differential equations in toroidal magnetic confinement geometries *Comput. Phys. Commun.* **181**(4) 715-19

[10] Jolliet S, McMillan B F, Villard L, Vernay T, Angelino P, Tran T M *et al.* 2012 Parallel filtering in global gyrokinetic simulations *J. Comput. Phys.* **231**(3) 745-58

[11] Tran T M, Appert K, Fivaz M, Jost G, Vaclavik J and Villard L 1999 Global gyrokinetic simulation of ion-temperature-gradient-driven instabilities using particles *Theory of Fusion Plasmas, Int. Workshop (Editrice Compositori, Bologna, Societa Italiana di Fisica)* **45**

[12] Jolliet S, Bottino A, Angelino P, Hatzky R, Tran T M, McMillan B F *et al.* 2007 A global collisionless PIC code in magnetic coordinates *Comput. Phys. Commun.* **177**(5) 409-25

[13] Biancalani A, Bottino A, Lauber P and Zarzoso D 2014 Numerical validation of the electromagnetic gyrokinetic code NEMORB on global axisymmetric modes *Nucl. Fusion* 54(10) 104004

[14] Abel I G, Plunk G G, Wang E, Barnes M, Cowley S C, Dorland W and Schekochihin A A 2013 Multiscale gyrokinetics for rotating tokamak plasmas: fluctuations, transport and energy flows *Reports on Progress in Physics* **76**(11) 116201

[15] Grandgirard V, Brunetti M, Bertrand P, Besse N, Garbet X, Ghendrih P *et al.* 2006 A drift-kinetic semi-Lagrangian 4D code for ion turbulence simulation *J. Comput. Phys.* **217**(2) 395-423

[16] Idomura Y, Tokuda S and Wakatani M 1999 Gyrokinetic theory of slab ion temperature gradient mode in negative shear tokamaks *Phys. Plasmas* **6**(12) 4658-4671

Journal of Physics: Conference Series **775** (2016) 012011 doi:10.1088/1742-6596/775/1/012011

Edge rotation from momentum transport by neutrals

JT Omotani[1], SL Newton[1,2], I Pusztai[1] and T Fülöp[1]
[1]Department of Physics, Chalmers University of Technology, 41296 Gothenburg, Sweden
[2]CCFE, Culham Science Centre, Abingdon, Oxon, OX14 3DB, UK

E-mail: omotani@chalmers.se

Abstract. Due to their high cross field mobility, neutral atoms can have a strong effect on transport even at the low relative densities found inside the separatrix. We use a charge-exchange dominated model for the neutrals, coupled to neoclassical ions, to calculate momentum transport when it is dominated by the neutrals. We can then calculate self-consistently the radial electric field and predict the intrinsic rotation in an otherwise torque-free plasma. Using a numerical solver for the ion distribution to allow arbitrary collisionality, we investigate the effects of inverse aspect ratio and elongation on plasma rotation. We also calculate the rotation of a trace carbon impurity, to facilitate future comparison to experiments using charge exchange recombination spectroscopy diagnostics.

1. Introduction

Tokamak plasmas typically rotate, even in the absence of external momentum input. Rotation is crucial to performance since it stabilizes magnetohydrodynamic instabilities such as the resistive wall mode [1] and flow shear suppresses turbulence in the pedestal, leading to the high-confinement mode (H-mode) [2]. Intrinsic rotation is particularly important for future tokamaks such as ITER because the torque from NBI heating will be weaker in larger devices. Neutral particles, which are always present in the edge of the confined plasma volume, provide a mechanism to generate intrinsic rotation. They may give large momentum transport even at low relative density due to their high cross field mobility, as has been demonstrated theoretically [3–11]. The importance of neutrals for the edge plasma is confirmed by experimental evidence of the influence of neutral density and location on H-mode confinement and the low to high confinement transition threshold [12–22]. The neutrals may be localized poloidally due to gas fuelling at a particular location, or divertor recycling causing the neutrals to be concentrated near the X-point. The resulting momentum transport and rotation is then controlled by the location of the neutrals [10, 11, 23]. In Ref. [23] we presented a numerical framework with which the rotation can be calculated for such scenarios, allowing for arbitrary plasma collisionality. In this paper we validate our numerical procedure against the analytical limits given in Refs. [10, 11] for asymptotic collisionality regimes. We investigate the effects of the lowest order equilibrium shaping parameters (i.e. those with the smallest poloidal mode numbers), inverse aspect ratio ϵ and elongation κ. Finally, we consider impurity rotation. Plasma rotation is usually determined by charge exchange recombination spectroscopy [24] which often measures the rotation of an impurity species. The main ion and impurity rotation are known to differ [25]. We demonstrate that by including a trace impurity species in our simulations, both the bulk ion and the impurity rotation can be calculated within our framework for arbitrary collisionality of either species.

2. Momentum transport by neutrals

2.1. Model

The evolution of the radial electric field E_r in a tokamak plasma is determined by the radial current through Ampère's law. In steady state the radial current must vanish. The radial particle fluxes $\mathbf{\Gamma}_a$ of the charged species are then determined by the toroidal component of the momentum equations for species a [26],

$$\langle e_a \mathbf{\Gamma}_a \cdot \nabla\psi \rangle = \left\langle R\hat{\boldsymbol{\zeta}} \cdot \nabla \cdot \boldsymbol{\pi}_a \right\rangle - \langle RF_{a\zeta} \rangle - \left\langle n_a e_a R E_\zeta^{(A)} \right\rangle \tag{1}$$

for ions, electrons or neutrals. Here $\langle ... \rangle$ denotes the flux surface average, $2\pi\psi$ is the poloidal flux, R is the major radius, $\hat{\boldsymbol{\zeta}} = \nabla\zeta / |\nabla\zeta|$ is the unit vector in the direction of the toroidal angle ζ, $\boldsymbol{E}^{(A)}$ is the induced electric field and for the species a, e_a is the charge, $\boldsymbol{\pi}_a$ the viscosity tensor and $\boldsymbol{F}_a = \sum_b \boldsymbol{F}_{ab}$ the total friction force due to collisions with all other species b. $E_\zeta^{(A)}$ does not contribute to the radial current due to quasineutrality and $\sum_a \boldsymbol{F}_a = 0$ as momentum is conserved in collisions. For charged particles the viscosity component in (1) is typically negligible [26], leading to automatic ambipolarity at lowest order in the gyroradius expansion of standard neoclassical theory for a fully ionized plasma, so the radial electric field $E_r = -|\nabla\psi| \partial\Phi_0/\partial\psi$ is not constrained at this order [27–29]. However, neutrals have a non-negligible viscosity due to their high radial mobility which, as has been shown in the previous analytical work and is outlined below, depends on the toroidal velocity directly and therefore can constrain the electric field.

Since $\nabla(R\hat{\boldsymbol{\zeta}})$ is an antisymmetric tensor, while $\boldsymbol{\pi}_a$ is symmetric, then in the absence of external torques the vanishing of the radial current in steady state, where the sum on species a is over ions, electrons and neutrals, implies

$$0 = j_r = \sum_a \langle e_a \mathbf{\Gamma}_a \cdot \nabla\psi \rangle \approx \left\langle \nabla \cdot \left(R\hat{\boldsymbol{\zeta}} \cdot \boldsymbol{\pi}_{\mathrm{n}} \right) \right\rangle = \frac{1}{V'} \frac{d}{d\psi} \left(V' \left\langle R\hat{\boldsymbol{\zeta}} \cdot \boldsymbol{\pi}_{\mathrm{n}} \cdot \nabla\psi \right\rangle \right)$$

$$\Rightarrow \left\langle R\hat{\boldsymbol{\zeta}} \cdot \boldsymbol{\pi}_{\mathrm{n}} \cdot \nabla\psi \right\rangle = 0, \tag{2}$$

with $V' = \oint d\theta \, (\boldsymbol{B} \cdot \nabla\theta)^{-1}$, and we have noted that in the absence of external torque the boundary condition in the interior of the plasma is that the radial flux of toroidal angular momentum vanishes.

To compute the viscosity $\boldsymbol{\pi}_{\mathrm{n}}$ of the neutrals (denoted by subscript n) we solve the steady state neutral kinetic equation [6], assuming charge exchange (CX) dominates so that we may neglect other atomic processes,

$$\boldsymbol{v} \cdot \nabla f_{\mathrm{n}} = \frac{1}{\tau_{\mathrm{CX}}} \left(\frac{n_{\mathrm{n}}}{n_{\mathrm{i}}} f_{\mathrm{i}} - f_{\mathrm{n}} \right), \tag{3}$$

where f_a and n_a are the distribution function and density of a species a (we denote the bulk ions by subscript i). The CX collision frequency is $\tau_{\mathrm{CX}}^{-1} = n_{\mathrm{i}} \langle \sigma v \rangle_{\mathrm{CX}} \simeq 2.93 n_{\mathrm{i}} \sigma_{\mathrm{CX}} (T_{\mathrm{i}}/m_{\mathrm{i}})^{1/2}$ [6], which is larger than the ionization or recombination rates for tokamak edge parameters [30]; the CX rate is $\sim 40\%$ larger than the ionization rate at the 300 eV temperature we consider below, while recombination is negligible above a few eV [31]. Since we assume that the neutrals are dominated by charge exchange collisions with the ions, their temperature is the same as the ion temperature to lowest order, as the solution below shows. $\langle \sigma v \rangle_{\mathrm{CX}}$ is the thermal charge exchange rate coefficient and σ_{CX} is the charge exchange cross section for thermal particles. We will assume that the relative neutral density $n_{\mathrm{n}}/n_{\mathrm{i}}$ is small enough that the ion distribution is not affected directly by the neutrals, i.e. f_{i} is independent of f_{n}. We can then solve (3) with a

Journal of Physics: Conference Series **775** (2016) 012011 doi:10.1088/1742-6596/775/1/012011

short mean-free-path expansion $f_{\rm n} = f_{\rm n}^{(0)} + f_{\rm n}^{(1)} + ...$ in $\tau_{\rm CX} v_{\rm T}/L \ll 1$, where $v_{\rm T} = \sqrt{2T_{\rm i}/m_{\rm i}}$ is the thermal velocity, with the ion temperature $T_{\rm i}$ and mass $m_{\rm i}$, and L is a typical length scale of plasma parameters, giving

$$f_{\rm n}^{(0)} = \frac{n_{\rm n}}{n_{\rm i}} f_{\rm i}; \quad f_{\rm n}^{(1)} = -\tau_{\rm CX} \boldsymbol{v} \cdot \nabla f_{\rm n}^{(0)} = -\tau_{\rm CX} \boldsymbol{v} \cdot \nabla \left(\frac{n_{\rm n}}{n_{\rm i}} f_{\rm i} \right). \tag{4}$$

The neutral viscosity is then

$$\boldsymbol{\pi}_{\rm n} = m_{\rm i} \int d^3 v \left(\boldsymbol{v}\boldsymbol{v} - \frac{1}{3} v^2 \mathbb{I} \right) f_{\rm n} = \frac{n_{\rm n}}{n_{\rm i}} \boldsymbol{\pi}_{\rm i} - \tau_{\rm CX} m_{\rm i} \int d^3 v \left(\boldsymbol{v}\boldsymbol{v} - \frac{1}{3} v^2 \mathbb{I} \right) \boldsymbol{v} \cdot \nabla \left(\frac{n_{\rm n}}{n_{\rm i}} f_{\rm i} \right). \tag{5}$$

The first term can be neglected as the off-diagonal components of $\boldsymbol{\pi}_{\rm i}$ are small, leaving

$$\begin{aligned} R\hat{\boldsymbol{\zeta}} \cdot \boldsymbol{\pi}_{\rm n} \cdot \nabla\psi &\approx -R\tau_{\rm CX} m_{\rm i} \int d^3 v \left(\hat{\boldsymbol{\zeta}} \cdot \boldsymbol{v} \right) \left(\nabla\psi \cdot \boldsymbol{v} \right) \boldsymbol{v} \cdot \nabla \left(\frac{n_{\rm n}}{n_{\rm i}} f_{\rm i} \right) \\ &\approx -\frac{R\tau_{\rm CX} m_{\rm i}}{n_{\rm i}} \frac{dn_{\rm n}}{d\psi} \int d^3 v \left(\hat{\boldsymbol{\zeta}} \cdot \boldsymbol{v} \right) \left(\nabla\psi \cdot \boldsymbol{v} \right)^2 f_{\rm i}, \end{aligned} \tag{6}$$

assuming that the radial gradient of $n_{\rm n}$ dominates over those of the background plasma profiles and over any poloidal gradients.

Accounting for the gyroradius correction, the ion distribution function at the particle position $\boldsymbol{r}$ is

$$f_{\rm i}(\boldsymbol{r}) \approx f_{\rm i,gc0}(\boldsymbol{r}) - \frac{e\Phi_1}{T_{\rm i}} f_{\rm i,gc0}(\boldsymbol{r}) - \boldsymbol{\rho} \cdot \nabla f_{\rm i,gc0}(\boldsymbol{r}) + g_{\rm i}(\boldsymbol{r}), \tag{7}$$

where $f_{\rm i,gc0}$ is a Maxwellian, $\boldsymbol{\rho} = \boldsymbol{r} - \boldsymbol{R}_{\rm gc}$ is the gyroradius vector with $\boldsymbol{R}_{\rm gc}$ the guiding centre position, $\Phi_1 = \Phi - \langle\Phi\rangle$ is the poloidally varying part of the electrostatic potential Φ and 0, 1 subscripts refer to the order in a $\delta = \rho/L$ gyroradius expansion. We will obtain the non-adiabatic part of the perturbed distribution function, $g_{\rm i} = f_{\rm i,gc1} + \frac{e_{\rm i}\Phi_1}{T_{\rm i}} f_{\rm i,gc0}$, from numerical solutions of the first order drift kinetic equation. The first two terms on the right of (7) are isotropic in velocity, so do not contribute to (6), and we have

$$R\hat{\boldsymbol{\zeta}} \cdot \boldsymbol{\pi}_{\rm n} \cdot \nabla\psi \approx \left(R\hat{\boldsymbol{\zeta}} \cdot \boldsymbol{\pi}_{\rm n} \cdot \nabla\psi \right)^{\rm (dia)} + \left(R\hat{\boldsymbol{\zeta}} \cdot \boldsymbol{\pi}_{\rm n} \cdot \nabla\psi \right)^{\rm (neo)}, \tag{8}$$

where

$$\left(R\hat{\boldsymbol{\zeta}} \cdot \boldsymbol{\pi}_{\rm n} \cdot \nabla\psi \right)^{\rm (dia)} = \frac{\tau_{\rm CX} R^4 B_{\rm p}^4 T_{\rm i}^2}{eB^2} \frac{dn_{\rm n}}{d\psi} \left(\frac{d\ln p_{\rm i}}{d\psi} + \frac{e}{T_{\rm i}} \frac{d\Phi_0}{d\psi} + \frac{d\ln T_{\rm i}}{d\psi} \right) \tag{9}$$

$$\left(R\hat{\boldsymbol{\zeta}} \cdot \boldsymbol{\pi}_{\rm n} \cdot \nabla\psi \right)^{\rm (neo)} = -\frac{\tau_{\rm CX} R^2 B_{\rm p}^2 I m_{\rm i}}{2Bn_{\rm i}} \frac{dn_{\rm n}}{d\psi} \int d^3 v \, v_\parallel v_\perp^2 g_{\rm i}, \tag{10}$$

with $B_{\rm p}$ the poloidal magnetic field strength, $I = R\boldsymbol{B} \cdot \hat{\boldsymbol{\zeta}}$ and parallel and perpendicular taken relative to the background magnetic field. Drift velocity terms do not appear in (10) as they are second order in δ. This momentum flux is first order in δ, and therefore large compared to that of the ions, as the neutral distribution $f_{\rm n}^{(1)}$ is strongly anisotropic in gyroangle.

2.2. Interpretation

It is useful to write the momentum flux (8) in terms of the toroidal velocity and heat flux in order to explain the physical origins of the intrinsic momentum flux. This is done as follows. We can write (9) in terms of the toroidal components of the diamagnetic flow and heat flux [11]

$$V_\zeta^{(\mathrm{dia})} = -\frac{RB_\mathrm{p}^2 T_\mathrm{i}}{eB^2}\left(\frac{d\ln p_\mathrm{i}}{d\psi} + \frac{e}{T_\mathrm{i}}\frac{d\Phi_0}{d\psi}\right); \quad \frac{2q_\zeta^{(\mathrm{dia})}}{5p_\mathrm{i}} = -\frac{RB_\mathrm{p}^2 T_\mathrm{i}}{eB^2}\frac{d\ln T_\mathrm{i}}{d\psi}, \tag{11}$$

as

$$\left(R\hat{\boldsymbol{\zeta}}\cdot\boldsymbol{\pi}_\mathrm{n}\cdot\nabla\psi\right)^{(\mathrm{dia})} = -\tau_\mathrm{CX}R^3B_\mathrm{p}^2T_\mathrm{i}\frac{dn_\mathrm{n}}{d\psi}\left(V_\zeta^{(\mathrm{dia})} + \frac{2q_\zeta^{(\mathrm{dia})}}{5p_\mathrm{i}}\right). \tag{12}$$

To rewrite (10) note that $v_\| v_\perp^2 = \frac{2}{5}v_\| v^2 - \frac{2}{5}v^3 P_3(\xi)$ where $\xi = v_\|/v$ is the cosine of the pitch angle and $P_3(\xi) \equiv 5\xi^3/2 - 3\xi/2$ is the third order Legendre polynomial. The neoclassical part of the total heat flux is

$$\frac{I}{RB}\int d^3v\,\frac{m_\mathrm{i}}{2}v_\| v^2 g_\mathrm{i} \equiv \frac{I}{RB}Q_{\mathrm{i}\|}^{(\mathrm{neo})} \approx Q_{\mathrm{i}\zeta}^{(\mathrm{neo})} = q_{\mathrm{i}\zeta}^{(\mathrm{neo})} + \frac{5p_\mathrm{i}}{2}V_{\mathrm{i}\zeta}^{(\mathrm{neo})} + \mathcal{O}(\delta^3) \tag{13}$$

and so (10) gives

$$\left(R\hat{\boldsymbol{\zeta}}\cdot\boldsymbol{\pi}_\mathrm{n}\cdot\nabla\psi\right)^{(\mathrm{neo})} \approx -\tau_\mathrm{CX}R^3B_\mathrm{p}^2T_\mathrm{i}\frac{dn_\mathrm{n}}{d\psi}\left(V_{\mathrm{i}\zeta}^{(\mathrm{neo})} + \frac{2}{5p_\mathrm{i}}q_{\mathrm{i}\zeta}^{(\mathrm{neo})}\right) + \frac{\tau_\mathrm{CX}R^2B_\mathrm{p}^2 I m_\mathrm{i}}{5Bn_\mathrm{i}}\frac{dn_\mathrm{n}}{d\psi}\int d^3v\,v^3P_3(\xi)g_\mathrm{i}. \tag{14}$$

Thus we have finally

$$\left(R\hat{\boldsymbol{\zeta}}\cdot\boldsymbol{\pi}_\mathrm{n}\cdot\nabla\psi\right) \approx -\tau_\mathrm{CX}R^3B_\mathrm{p}^2T_\mathrm{i}\frac{dn_\mathrm{n}}{d\psi}\left(V_{\mathrm{i}\zeta} + \frac{2}{5p_\mathrm{i}}q_{\mathrm{i}\zeta}\right) + \frac{\tau_\mathrm{CX}R^2B_\mathrm{p}^2 I m_\mathrm{i}}{5Bn_\mathrm{i}}\frac{dn_\mathrm{n}}{d\psi}\int d^3v\,v^3P_3(\xi)g_\mathrm{i} \tag{15}$$

and we see that the steady state condition (2) is largely set by a balance between the toroidal friction and heat friction, with an additional contribution from the third Legendre component of the distribution function.[1]

The plasma flow and heat flux take the general form [26]

$$\boldsymbol{V}_\mathrm{i} = -\left(\frac{d\Phi_0}{d\psi} + \frac{1}{en_\mathrm{i}}\frac{dp_\mathrm{i}}{d\psi}\right)R\hat{\boldsymbol{\zeta}} - \frac{kI}{e\langle B^2\rangle}\frac{dT_\mathrm{i}}{d\psi}\boldsymbol{B} \tag{16}$$

$$\boldsymbol{q}_\mathrm{i} = -\frac{5p_\mathrm{i}}{2e}\frac{dT_\mathrm{i}}{d\psi}R\hat{\boldsymbol{\zeta}} + \frac{5lIp_\mathrm{i}}{2e\langle B^2\rangle}\frac{dT_\mathrm{i}}{d\psi}\boldsymbol{B}, \tag{17}$$

where k and l are dimensionless parameters that depend on the plasma collisionality. k varies from positive in the Pfirsch-Schlüter regime, (21), to negative in the banana regime, (24), while l remains positive, with $l = 1$ in the Pfirsch-Schlüter regime and slightly smaller in the banana regime, (24).

The neutral source is often poloidally localized, for example from a gas-puff valve or divertor recycling. Modelling the neutral density as a δ-function at $\theta = \theta_\mathrm{n}$, the flux surface average in (2) simply evaluates its argument at the poloidal location θ_n. Neglecting here, for illustrative

[1] Note that this form is evaluated for arbitrary bulk ion collisionality. The g_i appearing here corresponds to the usual $F + g^{\mathrm{banana}}$ in the low collisionality regime, giving rise to the equivalent expression (17) in Ref. [11].

purposes, the P_3 term we can use (15) in (2) to solve for the electric field, and hence also the rotation, as

$$-\frac{d\Phi_0}{d\psi} \approx \frac{T_\mathrm{i}}{en_\mathrm{i}}\frac{dn_\mathrm{i}}{d\psi} + \frac{1}{e}\left(2 + \frac{(k-l)I^2}{R_\mathrm{n}^2\langle B^2\rangle}\right)\frac{dT_\mathrm{i}}{d\psi}; \quad V_{\mathrm{i}\varsigma} \approx \left(R + \frac{(k-l)I^2R}{R_\mathrm{n}^2\langle B^2\rangle} - \frac{kI^2}{R\langle B^2\rangle}\right)\frac{1}{e}\frac{dT_\mathrm{i}}{d\psi}, \quad (18)$$

where $R_\mathrm{n} = R(\theta_\mathrm{n})$ is the major radius at which the neutrals are located. Thus the sign of $(k-l)$ determines the sign of the variation of E_r with R_n. At high collisionality $(k-l) < 0$ so E_r is more negative for small R_n, while at low collisionality $(k-l) > 0$ so E_r is more positive for small R_n. An accurate analytic form in the banana regime was given in Ref. [11], (22) here; it has the same $(k-l)$ prefactor for its R_n dependence, indicating that the preceding discussion does capture the origin of the trends. The toroidal velocity has a similar variation with R_n as it comes from substituting $d\Phi_0/d\psi$ into (16). These features can be identified in the numerical results shown in the following section.

3. Simulation results

As can be seen from Eq. (10) we need the ion guiding centre distribution function to determine the steady state radial electric field and toroidal flow. To calculate this we use the Pedestal and Edge Radially-global Fokker-Planck Evaluation of Collisional Transport (PERFECT) neoclassical solver [32], used here as a radially-local solver. Taking a guess for $d\Phi_0/d\psi$ we use the distribution function from PERFECT to calculate the neutral momentum flux, and iterate on $d\Phi_0/d\psi$ to find the value consistent with the steady state condition, Eq. 2. While experimental equilibria can be used, here we take model divertor magnetic geometries based on analytical solutions to the Grad-Shafranov equation [33], allowing trends in the rotation behaviour with shaping parameters to be identified. The geometry is parameterized in terms of the inverse aspect ratio ϵ, elongation κ and triangularity δ, as well as the major radius R_X and height Z_X of the X-point, along with the toroidal $\beta_\mathrm{t} = 2\mu_0\bar{p}/B_0^2$ (where $\bar{p}$ is the volume averaged pressure) and plasma current. Dimensional scales are set by the major radius of the plasma centre R_0 and the vacuum magnetic field at the centre B_0. We take the nominal scan parameters representative of ITER [33]: $R_0 = 6.2\,\mathrm{m}$, $B_0 = 5.3\,\mathrm{T}$, $\epsilon = 0.32$, $\kappa = 1.7$, $\delta = 0.33$, $R_\mathrm{X} = (1 - 1.1\delta\epsilon)R_0$, $Z_\mathrm{X} = -1.1\kappa\epsilon R_0$ and $\beta_\mathrm{t} = 0.05$. We allow the plasma current to vary to keep the safety factor q_{95} of the flux surface with normalized poloidal flux $\psi_\mathrm{N} = 0.95$ fixed to the value 2.84, corresponding to a plasma current of 15 MA in the baseline equilibrium. The neutrals are poloidally localized, represented by a delta function[2] in poloidal angle $n_\mathrm{n} = \hat{n}_\mathrm{n}(\psi)\delta(\theta - \theta_\mathrm{n})$, on the $\psi_\mathrm{N} = 0.95$ flux surface. The main ion species and neutrals are both deuterium and we do not need to solve the electron kinetic equation here due to the small mass ratio.

Below we show scans plotted at fixed collisionality. We use the collisionality $\hat{\nu} = \nu_\mathrm{ii}L_\parallel/v_\mathrm{T}$, with ν_ii the ion-ion collision frequency, calculated using the connection length $L_\parallel = \int_{\theta=0}^{\theta=2\pi} dl_\parallel/2\pi$, where $dl_\parallel$ is the line element parallel to $\boldsymbol{B}$, which reduces to $L_\parallel \approx qR_0$ in the large aspect ratio limit. The baseline value is $\nu_\mathrm{ii}L_\parallel/v_\mathrm{T} \approx 1.26$ corresponding to $T_\mathrm{i} = 300$ eV and $n_\mathrm{i} = 10^{20}$ m^{-3} in the baseline equilibrium. We set the collisionality by varying the density at fixed temperature and logarithmic temperature gradient. To give explicit rotation velocities and electric fields in physical units, we choose a scale length for the ion temperature gradient, at the outboard midplane, of $L_{T\mathrm{i}} = 10$ cm in the baseline equilibrium. In section 3.2 where we scan the equilibrium geometry, we hold fixed the gradient in poloidal flux at the value $\partial \ln T_\mathrm{i}/\partial\psi = -1/L_{T\mathrm{i}}|\nabla\psi| \approx 0.911\ \mathrm{T}^{-1}\,\mathrm{m}^{-2}$ corresponding to $L_{T\mathrm{i}} = 10$ cm in the baseline equilibrium. We set $\partial n_\mathrm{i}/\partial\psi = 0$ here. The density gradient $\partial n_\mathrm{i}/\partial\psi$ and electrostatic potential

[2] The delta function representation takes the poloidal width of the neutral density variation to be narrow compared to the minor radius, but the poloidal width should be understood to be large compared to the radial gradient scale length, so that poloidal derivatives may be neglected.

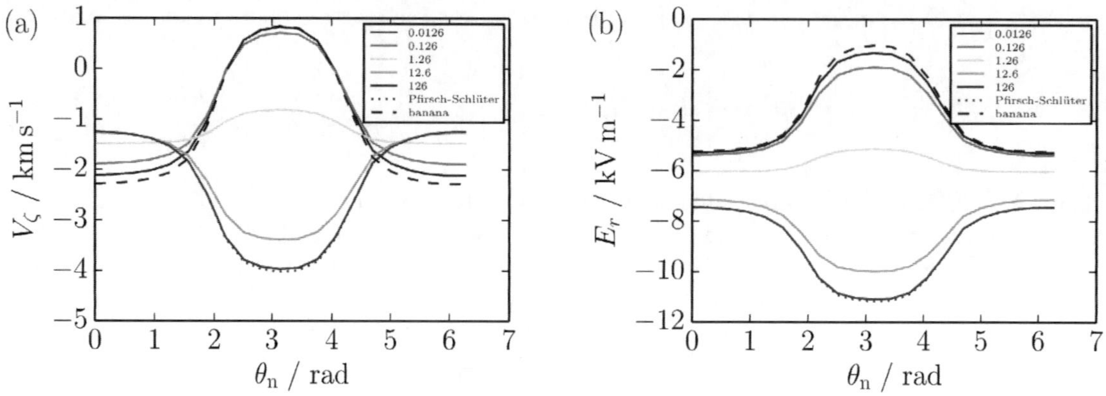

Figure 1. Comparison of analytical and numerical results for the dependence of (a) toroidal flow velocity and (b) radial electric field at the outboard midplane on poloidal location of the neutrals. Dashed lines show the banana regime analytic results, (23) and (22), and dotted lines the Pfirsch-Schlüter analytic results, (20) and (19). Solid lines show numerical scans, with colours from blue to red corresponding to low to high collisionality $\hat{\nu}$, as shown in the legend. A typical ion temperature scale length of 10 cm is taken, to remove the normalizations and give explicit rotation velocity and electric field values. $\theta_n = 0$ is the outboard midplane.

gradient $\partial\Phi_0/\partial\psi$ terms in the drift kinetic equation have identical velocity space structures. Therefore when we solve for $\partial\Phi_0/\partial\psi$ any density gradient gives an offset to $\partial\Phi_0/\partial\psi$ without affecting the flow on the flux surface we consider, as can be seen in Eq. (18).

3.1. Comparison to analytical theory

Analytical forms for the radial electric field and toroidal flow that result from momentum transport by neutrals were calculated in the Pfirsch-Schlüter and banana regimes in Ref. [11]. The electric field $-d\Phi_0/d\psi$ is a flux function, but $V_{i\zeta}$ varies poloidally and depends on both the major radius where the neutrals are localized, R_n, and the major radius where $V_{i\zeta}$ is evaluated, R. In the Pfirsch-Schlüter regime

$$-\frac{d\Phi_0}{d\psi}(R_n) = \frac{T_i}{en_i}\frac{dn_i}{d\psi} + \frac{1}{e}\frac{dT_i}{d\psi}\left(2 + \frac{(k-1)I^2}{\langle B^2\rangle R_n}\right) \tag{19}$$

$$V_{i\zeta}(R, R_n) = \frac{I^2}{e\langle B^2\rangle R}\frac{dT_i}{d\psi}\left(\frac{\langle B^2\rangle R^2}{I^2} - \frac{R^2}{R_n^2} + k\left(\frac{R^2}{R_n^2} - 1\right)\right) \tag{20}$$

and the coefficient k is

$$k = 1.97 + 0.09\frac{\langle B^2\rangle\left\langle\left(\nabla_\parallel \ln B\right)^2\right\rangle}{\left\langle\left(\nabla_\parallel B\right)^2\right\rangle}, \tag{21}$$

with the numerical prefactors corrected by a more accurate, numerical solution to the relevant Spitzer problems [34].

In the banana regime,

$$-\frac{d\Phi_0}{d\psi}(R_n) = \frac{T_i}{en_i}\frac{dn_i}{d\psi} + \frac{1}{e}\frac{dT_i}{d\psi}\left(2 + (k-l)\frac{f_2}{f_c}\frac{B_n I^2}{\langle B^2\rangle^{3/2} R_n^2}\right) \quad (22)$$

$$V_{i\zeta}(R, R_n) = \frac{I^2}{e\langle B^2\rangle R}\frac{dT_i}{d\psi}\left(\frac{\langle B^2\rangle R^2}{I^2} - k + (k-l)\frac{f_2}{f_c}\frac{R^2 B_n}{R_n^2\sqrt{\langle B^2\rangle}}\right), \quad (23)$$

where for a pure plasma

$$k = -1.17/(1 + 0.46 f_t/f_c); \quad l = 1/(1 + 0.46 f_t/f_c), \quad (24)$$

using an interpolation formula [26, 35] between the exact results for large and unit aspect ratio, with the usual definition of the fraction of circulating particles

$$f_c = 1 - f_t = \frac{3}{4}\langle B^2\rangle \int_0^{B_{max}^{-1}} d\lambda \frac{\lambda}{\langle\sqrt{1-\lambda B}\rangle}; \quad f_2 = \frac{15}{16}\langle B^2\rangle^{3/2}\int_0^{B_{max}^{-1}} d\lambda \frac{\lambda^2}{\langle\sqrt{1-\lambda B}\rangle}. \quad (25)$$

Fig. 1 shows the simulation results of a scan in the poloidal angle θ_n at which the neutrals are localized in the nominal geometry for a wide range of collisionalities. The analytical limits quoted above are also shown and good agreement can be seen at both high and low collisionality, verifying our numerical implementation. The trends are as expected from the discussion in Sec. 2.2.

3.2. Low order shaping parameters

In Ref. [23] we studied the effect on the electric field and rotation of varying the shaping parameters while keeping the neutrals localized at $\theta_n = \theta_X$, where θ_X is the poloidal angle of the X-point, representing a divertor recycling source or private flux region fuelling. We concluded that the shaping parameters most relevant for machine operation, triangularity and X-point position, only affect the electric field and rotation by changing the major radius where the neutrals are localized, for a fixed collisionality. Here we show that the same is true for the elongation κ, as long as we hold fixed the temperature gradient in poloidal flux space $dT_i/d\psi$ rather than the radial gradient dT_i/dr. Figs. 2(c,d) show that varying κ does not change the radial electric field or rotation, as expected since the major radius of the X-point does not vary with κ.

In contrast the response to varying the inverse aspect ratio ϵ does not follow the same trend with the major radius where the neutrals are localized. In Ref. [23] we found that the electric field increased with R_n at high collisionality and decreased at low collisionality. The radius of the X-point decreases with increasing ϵ, so Fig. 2(b) shows an opposite trend. In the unit aspect ratio limit, the mirror force on the inboard side becomes large and suppresses poloidal flow, so that the contributions parallel to $\boldsymbol{B}$ in (16) and (17) vanish in this limit. This leaves only the rigid rotation parts of the flow and heat flux, which are independent of collisionality. The electric field then approaches $-d\Phi_0/d\psi = (2/e)\,dT_i/d\psi \approx -0.547\ \mathrm{kV\,T^{-1}\,m^{-2}}$ (recall that $dn_i/d\psi = 0$ here) and the outboard midplane toroidal rotation approaches $V_{i\zeta} = (R/e)dT_i/d\psi \approx -3.39\ \mathrm{km\,s^{-1}}$, as can be seen in Figs. 2(a,b). In the large aspect ratio limit $\epsilon \to 0$, $f_t \to 0$ so $l \to 1$ in the banana regime (recall that $l = 1$ generally in the Pfirsch-Schlüter regime) while $\langle B^2\rangle \to I^2/R^2$ so we see from (17) that $q_{i,\zeta} \to 0$. As discussed in section 2.2 the heat flux gives rise to the intrinsic momentum flux that drives rotation, Eq. (15), and so the rotation also vanishes $V_{i\zeta} \to 0$ in this limit.[3]

[3] Again, we ought to account for the third Legendre harmonic component as well as the heat flux, but the accurate banana regime result (23) shows that the rotation indeed vanishes, since $f_2 \to 1$ while $R \approx R_n \approx R_0$ and $\sqrt{\langle B^2\rangle} \approx B_n \approx B_0$.

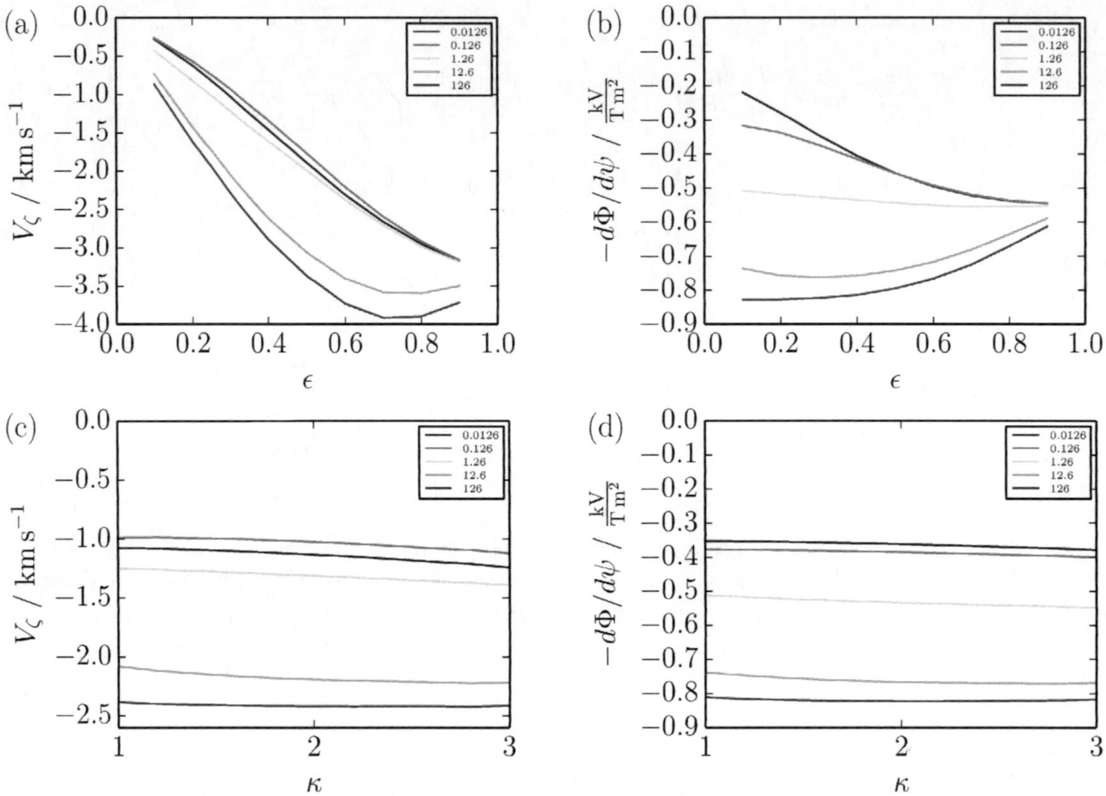

Figure 2. Response of the toroidal rotation $V_{i\zeta}$ and electric field $-d\Phi_0/d\psi$ at the outboard midplane to varying inverse aspect ratio ϵ (a,b) and elongation κ (c,d) for several collisionalities, with neutrals localized at the X-point. The shaping parameters not being scanned are kept at the baseline values. Colours blue to red again correspond to low to high collisionalities as shown in the legend. $dT_i/d\psi$ is held constant at the value corresponding to a gradient length scale of 10 cm in the baseline geometry.

3.3. Impurity rotation

Tokamak rotation is most often measured by charge exchange recombination spectroscopy [24]. Due to the high background of deuterium D_α radiation, an impurity species is most often used for the measurement. However, it is well known both theoretically [25] and experimentally [36] that impurity and bulk ion rotation may differ. For a trace impurity species, we may assume that the presence of the impurity does not affect the momentum transport through neutrals, and by including the impurity in our PERFECT simulations we can calculate the impurity rotation along with that of the main ions, for arbitrary collisionality of either species. The result is shown in Fig. 3(a) for a trace, fully ionized carbon impurity with density $n_z \approx 0.018 n_i$ where the neutrals are localized at the outboard midplane. Ref. [37] gives a convenient form in their Eq. (15) for the velocity difference between bulk ions and a collisional impurity, based on the results of Refs. [25, 38]. Here we neglect all $\mathcal{O}(Z^{-1})$ terms for consistency, where Z is the atomic number of the impurity. These analytical results for collisional and mixed collisionality (banana bulk ion and Pfirsch-Schlüter impurity) limits are compared to the simulation results in Fig. 3(b). The trends agree well, the $\sim 30\%$ discrepancy is consistent with $\mathcal{O}(Z^{-1})$ errors for carbon and the approximate collision operators used analytically. The analytical results show that the velocity difference is independent of the radial electric field, and therefore independent of the location

Journal of Physics: Conference Series **775** (2016) 012011 doi:10.1088/1742-6596/775/1/012011

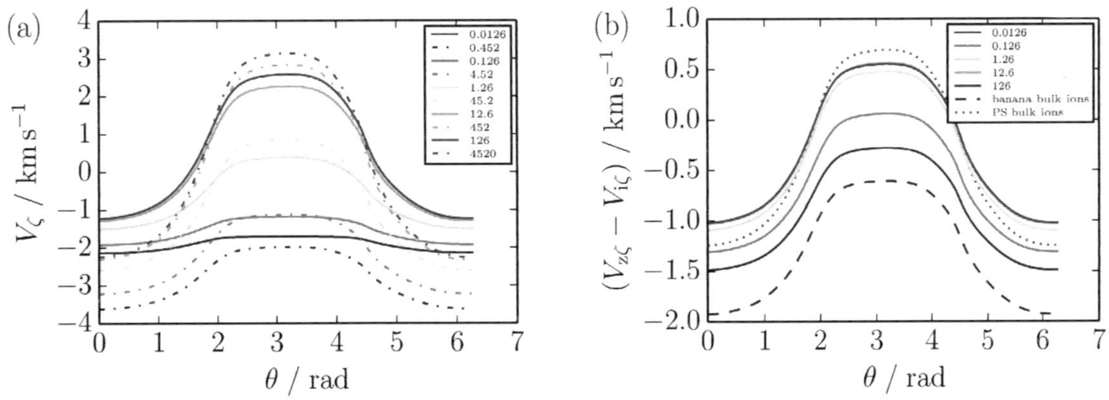

Figure 3. Poloidal variation of (a) the toroidal velocity of bulk ions (solid) and trace, fully ionized carbon impurity (dash-dotted) for neutrals localized at the outboard midplane $\theta_n = 0$ and (b) the difference between carbon and bulk ion velocities (solid), compared to analytical limits with collisional impurity and collisional (dotted) or banana regime (dashed) bulk ions. Colours from blue to red correspond to low to high collisionality $\hat{\nu}$, as shown in the legend. In (b), only the bulk ion collisionality is quoted. The ion temperature scale length is again taken to be 10 cm to produce explicit values. $\theta = 0$ is the outboard midplane.

of the neutrals in our framework, and indeed Fig. 3(b) is identical for simulations with different neutral locations. The quantitative differences highlight the importance of accurate numerical results for the interpretation of experimental impurity rotation measurements.

4. Conclusions

By considering the transport of toroidal momentum through charge-exchanging neutrals, we calculate the radial electric field and intrinsic rotation in a tokamak without external torque, where the flows are subsonic. We have compared the results of our numerical framework first presented in Ref. [23] to the analytical limits of Ref. [11] (with more accurate collisional flow coefficients) finding good agreement in both collisional and collisionless limits. We have extended the investigation of Ref. [23] to lower order shaping parameters, finding that the elongation κ has little effect on the electric field and rotation due to neutrals. In contrast, varying the inverse aspect ratio ϵ has more fundamental effects, for instance changing the mirror forces, resulting in trends in electric field and rotation that are not captured just by the collisionality and the major radius where the neutrals are localized, R_n. Finally, in order to facilitate comparisons to experimental measurements which usually obtain impurity rotation by charge exchange recombination spectroscopy, we have demonstrated that impurities can be included in the framework. Here we neglect their interactions with the neutrals, so impurities may be present only in trace quantities, but arbitrary collisionality of either ion species can be treated.

Acknowledgements

The authors are grateful to Matt Landreman for advice and help with the PERFECT code and to Stuart Henderson for assistance with the ADAS database. This work was supported by the Framework grant for Strategic Energy Research (Dnr. 2014-5392) and the International Career Grant (Dnr. 330-2014-6313) from Vetenskapsrådet.

References

[1] Hender T *et al.* 2007 *Nucl. Fusion* **47** S128

[2] Terry P W 2000 *Rev. Mod. Phys.* **72**(1) 109
[3] Hazeltine R, Calvin M, Valanju P and Solano E 1992 *Nucl. Fusion* **32** 3
[4] Catto P J 1994 *Phys. Plasmas* **1** 1936
[5] Helander P, Krasheninnikov S I and Catto P J 1994 *Phys. Plasmas* **1** 3174
[6] Catto P J *et al.* 1998 *Phys. Plasmas* **5** 3961
[7] Fülöp T *et al.* 1998 *Phys. Plasmas* **5** 3398
[8] Fülöp T *et al.* 1998 *Phys. Plasmas* **5** 3969
[9] Fülöp T *et al.* 2001 *Phys. Plasmas* **8** 5214
[10] Fülöp T *et al.* 2002 *Phys. Rev. Lett.* **89**(22) 225003
[11] Helander P *et al.* 2003 *Phys. Plasmas* **10** 4396
[12] Carreras B A *et al.* 1998 *Phys. Plasmas* **5** 2623
[13] Owen L W *et al.* 1998 *Plasma Phys. Controlled Fusion* **40** 717
[14] Gohil P *et al.* 2001 *Phys. Rev. Lett.* **86**(4) 644
[15] Boivin R L *et al.* 2000 *Phys. Plasmas* **7** 1919
[16] Field A R *et al.* 2002 *Plasma Phys. Controlled Fusion* **44** A113
[17] Valovic M *et al.* 2002 *Plasma Phys. Controlled Fusion* **44** A175
[18] Fukuda T *et al.* 2000 *Plasma Phys. Controlled Fusion* **42** A289
[19] Field A R *et al.* 2004 *Plasma Phys. Controlled Fusion* **46** 981
[20] Maingi R *et al.* 2004 *Plasma Phys. Controlled Fusion* **46** A305
[21] Joffrin E *et al.* 2014 *25th IAEA Fusion Energy Conference (FEC 2014)* EX/P5-40
[22] Tamain P *et al.* 2015 *J. Nucl. Mater.* **463** 450
[23] Omotani J *et al.* 2016 *Nuclear Fusion* **56** 124002
[24] Isler R C 1994 *Plasma Phys. Controlled Fusion* **36** 171
[25] Kim Y B, Diamond P H and Groebner R J 1991 *Phys. Fluids B* **3** 2050
[26] Helander P and Sigmar D J 2002 *Collisional transport in magnetized plasmas*
[27] Rosenbluth M N *et al.* 1971 *Plasma Phys. Controlled Nucl. Fusion Res., Volume I* p 495
[28] Hazeltine R D 1974 *Phys. Fluids* **17** 961
[29] Catto P J and Simakov A N 2005 *Phys. Plasmas* **12** 012501
[30] Helander P 1999 *Proceedings of the Joint Varenna-Lausanne Workshop* ed Connor J W, Sindoni E and Vaclavik J (Bologna) p 373
[31] Summers H P 2004 *The ADAS User Manual, v. 2.6* (http://www.adas.ac.uk)
[32] Landreman M *et al.* 2014 *Plasma Phys. Controlled Fusion* **56** 045005
[33] Cerfon A J and Freidberg J P 2010 *Phys. Plasmas* **17** 032502
[34] Landreman M 2013 *private communication*
[35] Taguchi M 1988 *Plasma Phys. Controlled Fusion* **30** 1897
[36] Kim J *et al.* 1994 *Phys. Rev. Lett.* **72**(14) 2199
[37] Catto P J and Simakov A N 2006 *Phys. Plasmas* **13** 052507
[38] Hirshman S and Sigmar D 1981 *Nucl. Fusion* **21** 1079

Journal of Physics: Conference Series **775** (2016) 012012 doi:10.1088/1742-6596/775/1/012012

Towards nonaxisymmetry; initial results using the Flux Coordinate Independent method in BOUT++

B W Shanahan, P Hill, B D Dudson

York Plasma Institute, Department of Physics, University of York, Heslington, York YO10 5DD, UK

E-mail: bws502@york.ac.uk

Abstract. Fluid simulation of stellarator edge transport is difficult due to the complexities of mesh generation; the stochastic edge and strong nonaxisymmetry inhibit the use of field aligned coordinate systems. The recent implementation of the Flux Coordinate Independent method for calculating parallel derivatives in BOUT++ has allowed for more complex geometries. Here we present initial results of nonaxisymmetric diffusion modelling as a step towards stellarator turbulence modelling. We then present initial (non-turbulent) transport modelling using the FCI method and compare the results with analytical calculations. The prospects for future stellarator transport and turbulence modelling are discussed.

1. Stellarator and Nonaxisymmetric Modelling

Historically, neoclassical transport has been the dominant loss mechanism in stellarators [1]. Recent optimizations have allowed for the minimization of neoclassical losses, which has culminated in the design of the Wendelstein 7-X stellarator [2, 3]. As Wendelstein 7-X has been optimized for neoclassical transport, turbulent transport could potentially become comparable to neoclassical losses. As such, it is becoming increasingly important to simulate turbulence in nonaxisymmetric configurations to determine its role in comparison to neoclassical transport in order to optimize performance.

In the core of stellarators, the closed flux surfaces and low collisionality facilitate the use of gyrokinetic codes such as GENE [4], which is currently the only technique for simulating stellarator turbulence. However, due to the small scales simulated in gyrokinetics, the computation is quite expensive for experimentally relevant temporal and spatial scales. Additionally, GENE simulations are currently localized to single flux surfaces or flux-tube geometries.

The edge of stellarators, however, includes stochastic regions and magnetic islands, and edge modelling in stellarators is currently limited to magnetohydrodynamic transport modelling to determine the steady state profiles. This is done primarily using EMC3 [5], which employs a Monte-Carlo solver to simulate three dimensional transport (non-turbulent) equations in order to determine steady state quantities for divertor profiles. The relatively high collisionality of edge plasmas both in stellarators and tokamaks justifies a fluid approach to turbulence simulations, however the current nature of plasma fluid turbulence simulations renders it difficult to simulate nonaxisymmetric magnetic geometries; stochastic regions are difficult in field-aligned coordinate systems.

Edge fluid turbulence modelling of tokamak plasmas often exploits the axisymmetry of tokamak configurations to reduce the computational expense. The nonaxisymmetric nature of stellarators, however, requires that simulations are fully three dimensional. For three dimensional tokamak modelling, it is often advantageous to align the computational grid to the magnetic field, which helps improve numerical efficiency. Typically, parallel dynamics exhibit a long wavelength, which allows for lower resolution in the parallel direction, and therefore faster computation. In stellarators, however, the complex magnetic geometry requires either a carefully designed field aligned system such as the unstructured grid used by the FINDIF code [6], or a nonaligned system since parallel nonuniformities are introduced along the magnetic field.

`BOUT++` was originally developed for flute reduced plasma models in field aligned geometries. Specifically, the three dimensions used were radial, toroidal, and parallel to the magnetic field. This system is inaccurate in stochastic regions and regions near magnetic X- and O-points; two coordinates (toroidal, field aligned) are parallel at X-points, causing numerical instability in the form of zero-volume elements. Recent work has sought to exploit techniques such as nonaligned coordinate systems which has allowed for X-point simulation in `BOUT++` [7, 8, 9]. We will show here that it is possible to simulate stellarator geometries in `BOUT++` using a non-field-aligned grid through the implementation of the Flux Coordinate Independent (FCI) method for parallel derivatives.

2. The Flux Coordinate Independent Method

The advantage of field aligned coordinate systems is that parallel derivatives are simplified to be taken along one dimension of the coordinate system, which is computationally efficient since the majority of turbulence models are described by separating perpendicular and parallel derivatives. One disadvantage of this method is that complex geometries such as magnetic nulls are poorly described and susceptible to numerical instabilities. A second disadvantage of field-aligned coordinate systems is the difficulty associated with generating the mesh for nonaxisymmetric fields. In most turbulence codes, field aligned grids are generated using a two dimensional poloidal cross section, and an assumption of axisymmetry which leads to a two dimensional equilibrium. Additionally, the presence of magnetic islands and stochastic magnetic field regions in stellarators render this method impractical.

Recently the Flux Coordinate Independent method for calculating parallel derivatives [10, 11] has been implemented in `BOUT++`. This method for calculating parallel derivatives has been implemented in other codes [12, 13], and is intuitively straightforward, as described in Figure 1.

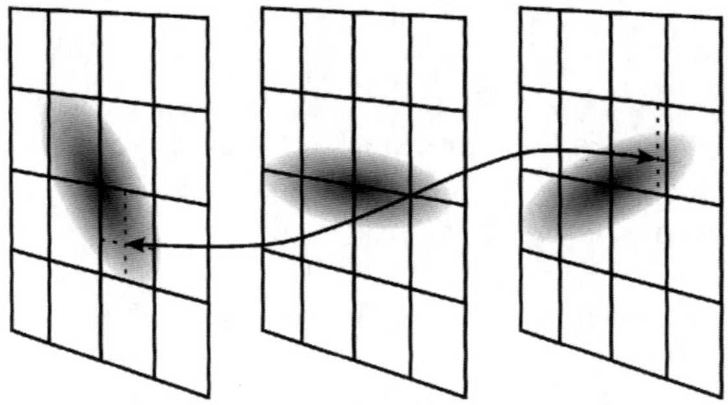

Figure 1. Illustration of Flux Coordinate Independent method for calculating parallel derivatives.

Based on the form of the magnetic field at a given point, the field line is followed to the next poloidal (or azimuthal) plane. The position at which the field line hits the next plane is interpolated to the nearest grid points, and a value for a given quantity is assigned based on that interpolation. This process is repeated on the previous plane, and a differential is calculated based on central differencing. As the FCI method is used solely for calculation of parallel derivatives, the process is independent of the poloidal grid configuration. In the work presented here, the poloidal grids are chosen to be Cartesian.

While this process is intuitively straightforward, it allows for more complex magnetic field configurations in comparison to structured grids. Previous work has used this model to simulate turbulence in the region of magnetic islands [14], verifying its suitability for magnetic X- and O-points. In the following section, the recent implementation of the FCI method is tested to determine if nonaxisymmetric modelling is possible within BOUT++.

3. Foundations for stellarator modelling in BOUT++

The FCI method has been implemented into BOUT++ and tested via the method of manufactured solutions [10], which determined that the operators converge to second order, as is expected for the central difference schemes in use. One of the aims of the research presented here is to provide the components necessary to simulate stellarator turbulence cases and evaluate the computational and developmental work required. The following subsection details the progress towards stellarator turbulence using the FCI method by describing the recent nonaxisymmetric test scenarios which have been implemented.

3.1. Diffusion test cases

To determine the efficacy of the FCI method as a tool for stellarator turbulence modelling, a test case of an infinite aspect ratio classical stellarator was constructed. A theoretical linear device with 4 helical coils was considered, as shown in Figure 2. A Poincaré plot of this configuration was created to ensure the existence of closed flux surfaces, as shown in Figure 3.

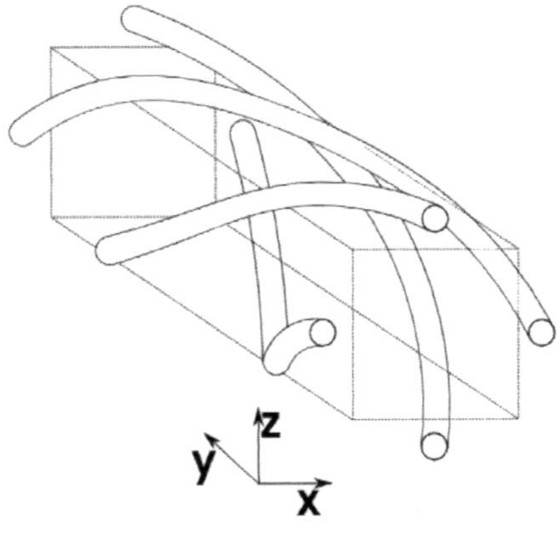

Figure 2. The straight stellarator test case; a very large aspect ratio classical stellarator showing helical coils and the inlaid Cartesian coordinate system (dashed).

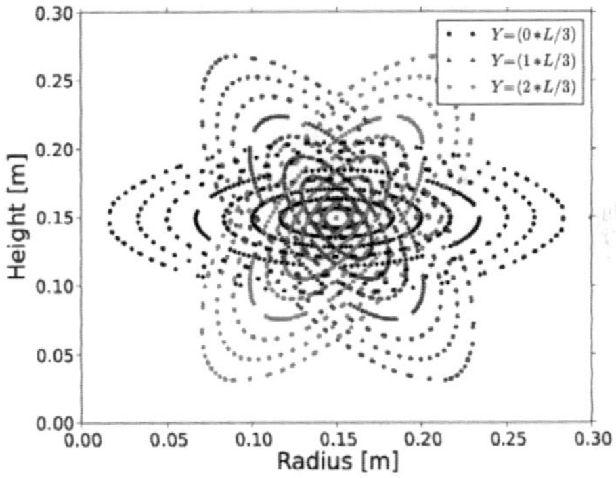

Figure 3. Poincaré plot of the straight stellarator indicating closed flux surfaces

This configuration has been implemented into the FCI grid generator [10], where it is possible to change several parameters including rotational transform, toroidal field and helical coil current/position. A typical grid with 16 Cartesian poloidal planes with a uniform 256 x 256 resolution can be generated in approximately 45 seconds.

As the FCI method is purely a tool for calculating parallel dynamics, a simple heat advection equation was modelled:

$$\frac{\partial f}{\partial t} = \nabla \cdot (\mathbf{bb} \cdot \nabla f) \equiv \nabla_{\parallel}^2 f \tag{1}$$

By solving Equation 1 for an initial three dimensional (non-field-aligned) Gaussian distribution of our test function f, and allowing the simulated to reach a saturated steady state, it is possible to trace out the flux surfaces (shown in Figure 3). The results of the diffusion simulation are shown in Figure 4.

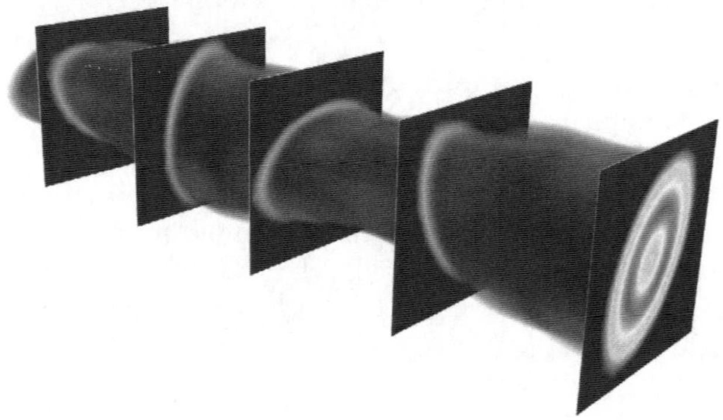

Figure 4. Heat diffusion in the straight stellarator test case; flux surfaces are correctly mapped out, qualitatively indicating proper calculation of parallel derivatives.

Figure 4 indicates that the FCI method for calculating parallel derivatives is correctly evaluating parallel dynamics; as an initial distribution is left to propagate along the field lines, the flux surfaces are traced out. The extent to which the interpolation of field lines in the perpendicular planes modifies the calculated quantities is tested in the following section.

3.2. Inherent Numerical Diffusion

There exists a strong anisotropy of heat conductivity in magnetized fusion plasmas. Parallel conductivity can be as high as 10^{10} times higher than perpendicular conductivity. It is therefore important to minimize errors in parallel operators, as even a small perpendicular pollution of parallel dynamics can lead to substantial errors [15]. One of the main sources of error for the FCI method is the interpolation, as every quantity is interpolated based on where the field lines intersect the next and previous perpendicular planes. The issue can be illustrated by considering a field aligned structure which is very small in the poloidal plane at a given grid point. Assuming the field line does not intersect a grid point in the next (or previous) perpendicular plane, the structure will be distributed between the four nearest grid points. This will then dissipate the function, causing a loss of accuracy and introducing an error. Currently, the FCI method in `BOUT++` utilizes cubic Hermite spline interpolation, but other methods are possible and can be used [12]. As a test of the interpolation, the previous diffusion case of a straight classical stellarator was implemented for 4 different grid resolutions. It is possible to determine the inherent numerical diffusion from interpolation by assuming that the diffusion follows the relation:

$$\frac{\partial f}{\partial t} = D\nabla_{\perp}^2 f = D\nabla \cdot (\nabla f - \mathbf{bb} \cdot \nabla f) \tag{2}$$

where D is the diffusion coefficient for the numerical diffusion of our test function. The inherent numerical diffusion from the interpolation scheme puts a limit on the minimum resolution which can be used, as low resolution grids will introduce a higher perpendicular numerical diffusion. Ideally, numerical perpendicular diffusion should be at least a factor of 10^{-8} smaller than the parallel dynamics [15]. The scaling of inherent perpendicular numerical diffusion coefficients with perpendicular mesh spacing in the straight stellarator geometry is shown in Figure 5, where the diffusion coefficients at a grid point just off-axis (r,z = 16,15cm) are normalized to the parallel diffusion. For this analysis, the number of parallel grid points was fixed at 16, and the number of perpendicular grid points varied; 64x64 resolution gives a mesh spacing of 4.76mm, 128x128 resolution gives a perpendicular mesh spacing of 2.29mm, 256x256 resolution indicates a mesh spacing of 1.16mm, and 512x512 resolution has a mesh spacing of 0.59mm. For reference, the domain size is always set to 30cm x 3m x 30cm, and the parallel resolution is chosen to be a constant 19.6cm. Figure 6 indicates the loss of our test function f due to numerical diffusion as a function of time for these various resolutions.

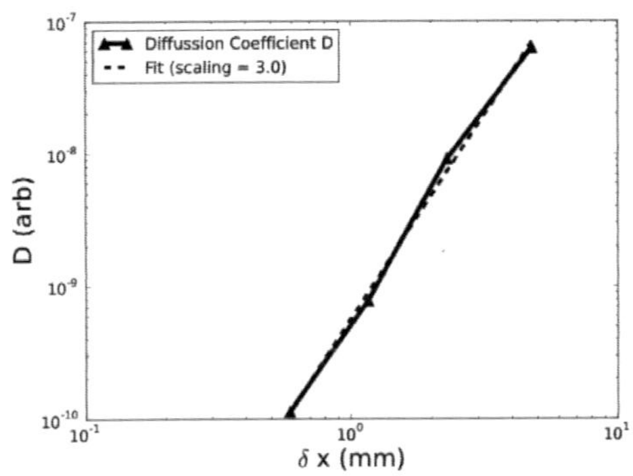

Figure 5. Inherent numerical perpendicular diffusion as a function of poloidal mesh spacing in the straight stellarator test case. The fit shows third order convergence, which is broadly in line with previous work [10].

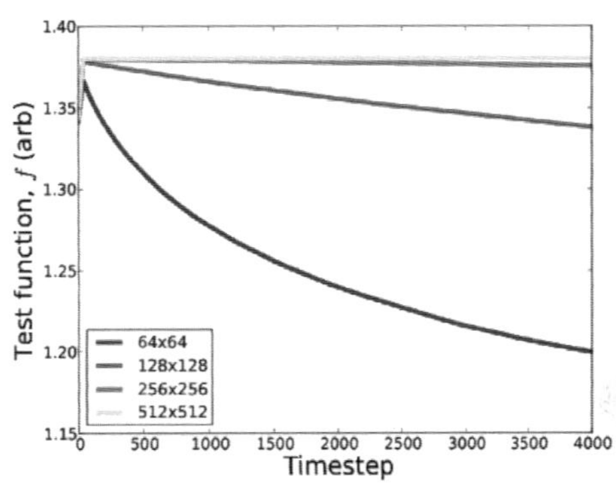

Figure 6. The test function f at the center of the domain as a function of time for various resolutions; inherent numerical diffusion serves as an artificial sink at lower resolutions.

From Figures 5 and 6 it appears that the optimal resolution for an FCI mesh is 256 by 256, as the perpendicular diffusion is at least 10^{-8} smaller than the parallel diffusion. Of course, higher resolution cases are more precise but are also more computationally expensive.

Having quantified the diffusion inherent in the FCI method for parallel derivatives within BOUT++, we arrive at a minimum resolution required for this nonaxisymmetric configuration, which is comparable to the resolution one would choose for a given turbulence case for a system of this size, as $\rho_s \approx 1$mm. We have therefore provided evidence that nonaxisymmetric modelling is possible in BOUT++, as the most difficult barrier to stellarator modelling is the ability to correctly capture parallel dynamics. The next section describes recent work in implementing a transport model which is a subset of the EMC3 model, which intends to test the efficacy of BOUT++ as an alternative to common methods used in stellarator modelling.

4. Transport modelling

EMC3 [5] is a commonly used tool to simulate the steady state profiles of stellarator edge plasmas, which allows for the reconstruction of heat flux profiles for divertor optimization. Recent work [16] has looked to test the EMC3 model against analytic solutions of the one dimensional transport model. Here, we present the first results following this work in BOUT++ to determine if BOUT++ can effectively and efficiently solve a simplified form of the EMC3 equations. The EMC3 equations can be reduced to a one dimensional model which captures isothermal parallel dynamics, which are shown in [16] to be:

$$\frac{\partial n}{\partial t} = -\nabla \cdot (nv_{\|}) + S_i \tag{3}$$

$$\frac{\partial nv_{\|}}{\partial t} = -v_{\|}\nabla_{\|}nv_{\|} - 2T_{e0}\nabla_{\|}n \tag{4}$$

where T_{e0} is the isothermal electron temperature, assumed here to be 5eV, n is the density, S_i is the constant source function, and $v_{\|}$ is the parallel velocity. To ensure that BOUT++ is finding the correct solution, the simulated steady state of the above transport model compared with an analytical solution, which has been previously found to be [16]:

$$n_i(x) = \frac{S_i x}{v_{\|}(x)c_s} \tag{5}$$

$$v_{\|}(x) = \frac{L}{2x} - \sqrt{\frac{L^2}{4x^2} - 1} \tag{6}$$

where S_i is again the ion source which is independent of position, x is the distance along the field line, c_s is the sound speed and L is the length of the domain, which spans from $-L/2$ to $L/2$. For all of the following simulations, the boundary conditions were implemented in accordance with [16]. Specifically, velocities were set to $\pm c_s$ (which is normalized to 1) at the upper and lower boundary, respectively. Densities were set to $\frac{S_i L}{2c_s}$.

The first implementation of this model into BOUT++ was done without the use of the FCI method, allowing for testing using conventional operators. Figures 7 and 8 illustrate a comparison of the analytical model and a simulation using finite volume (flux conserving) operators.

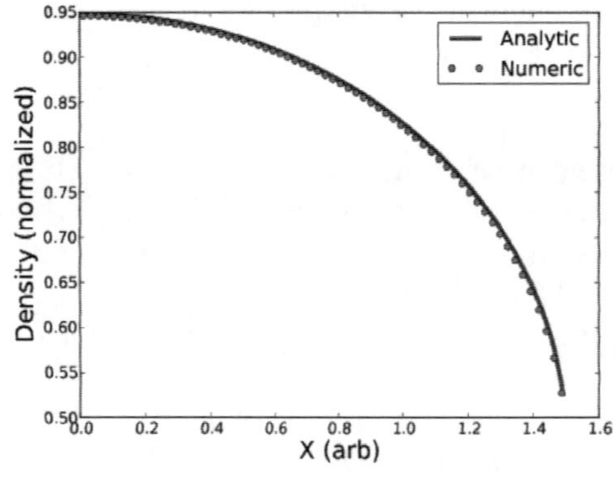

Figure 7. Steady state density in a 1D transport model and the analytical solution using conventional finite volume operators in BOUT++.

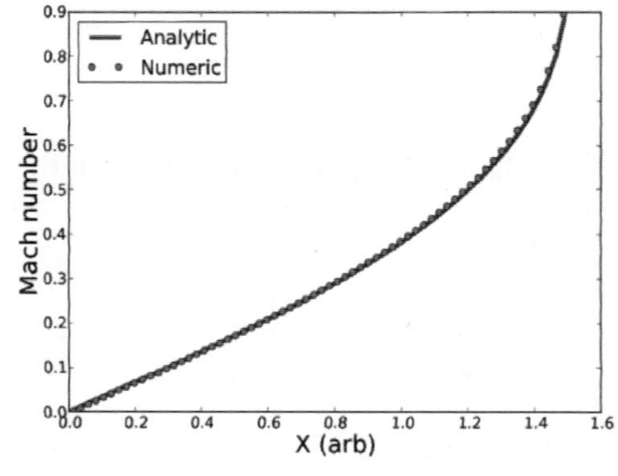

Figure 8. Comparison of the steady state velocity in a 1D transport model and an analytical solution using conventional finite volume operators in BOUT++.

It is clear from Figures 7 and 8 that BOUT++ is capable of simulating the correct profiles in these one dimensional transport equations. However, the FCI method utilizes a finite difference scheme, which is not conservative and therefore could introduce losses. As a test,

Journal of Physics: Conference Series **775** (2016) 012012 doi:10.1088/1742-6596/775/1/012012

the conventional (non-FCI) central differencing schemes were implemented and again compared to analytical solution, Figures 9 and 10

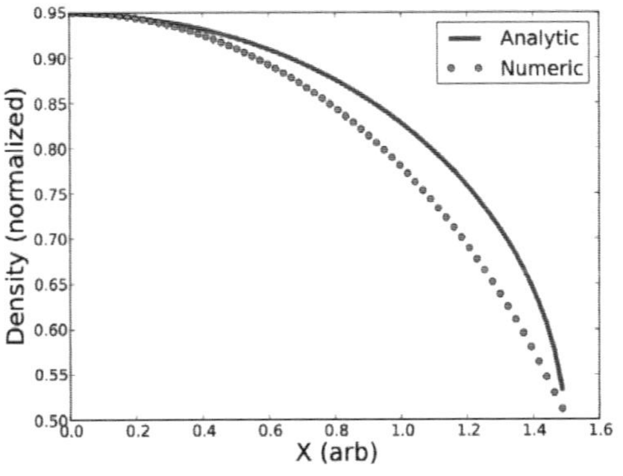

Figure 9. Steady state density in a 1D transport model and the analytical solution, having utilized a finite difference scheme within BOUT++.

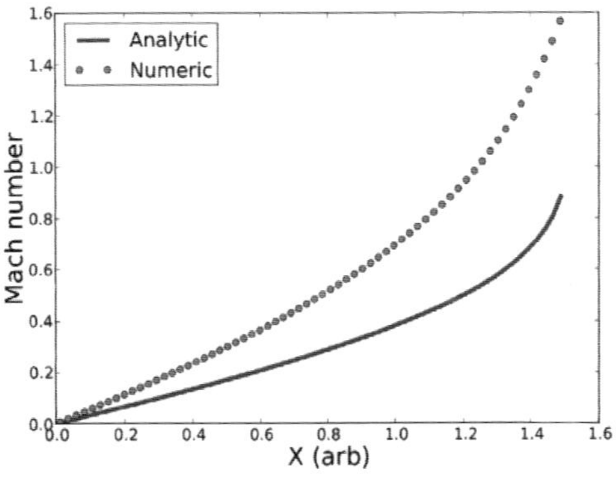

Figure 10. Comparison of the steady state velocity using conventional finite difference schemes in a transport model and an analytical solution.

Figures 9 and 10 indicate that the more simple central differencing scheme fails to reproduce the analytical solution, as the numerical steady state converges to incorrect profiles. Using this method, it is possible that quantities can be lost from the simulation, as these operators are non-conservative; flux exiting one computational cell is not necessarily entering the next. As such, it is important to minimize these losses when using a finite difference scheme, such as the FCI method. A method for improving these finite difference schemes will be discussed shortly.

To further test the Flux Coordinate Independent method for parallel derivatives, this one dimensional transport model was implemented in the geometry shown in Figure 11. This geometry has a completely straight field at the center of the domain, where the FCI method must not interpolate, and an increasingly strong helical field at larger minor radii where interpolation is essential. Furthermore, the magnetic field line length is nonuniform which (referring to Equation 5) creates a radially varying profile for density.

The one dimensional transport model was implemented into the geometry shown in Figure 11 at various resolutions. Similar to the straight stellarator test case, the parallel resolution (**y**) was held constant while the resolution of Cartesian poloidal planes was varied. To minimize losses, the FCI operators were modified to calculate derivatives based on the flux at the at the grid cell faces; the flux at the edges of each computational cell was averaged with the flux at the edge of its neighboring cells. While this still does not guarantee the conservation properties of finite volume methods, the quantities are more closely conserved than in simple finite difference schemes. Figures 12 and 13 illustrate the results of the transport simulation at the center of the domain shown in Figure 11 for three different poloidal resolutions. As these plots are taken at the center of the domain where the field line is straight, they can be compared to the previous finite difference results, Figures 9 and 10, and a clear improvement is seen.

As the simulations correctly reproduces the behavior of the analytic model at the center of the domain, it can be concluded that the losses due to the central differencing scheme used by FCI have been reduced. Furthermore, there is no dependence on poloidal resolution as the interpolation scheme is not used on straight field lines. To test the interpolation of the FCI method, it is useful to examine the results away from the center where the field lines are helical, as shown in Figures 14 and 15. Specifically, these results were taken at about two-thirds of the

Journal of Physics: Conference Series **775** (2016) 012012 doi:10.1088/1742-6596/775/1/012012

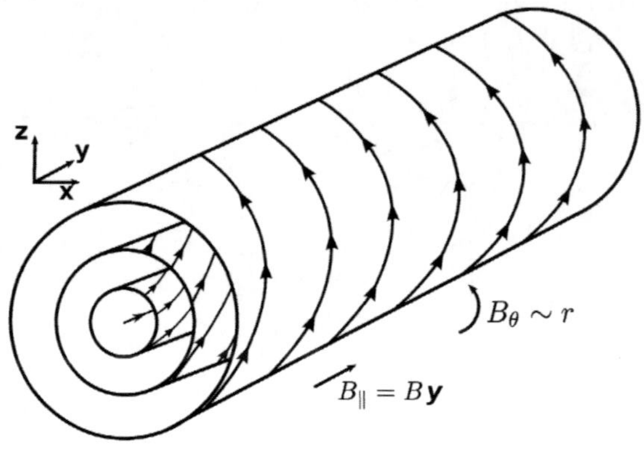

Figure 11. The geometry used to test the FCI method when solving the one dimensional transport model. The azimuthal magnetic field is proportional to the minor radius r, allowing a test of straight field lines on axis, and a test of interpolation away from the center.

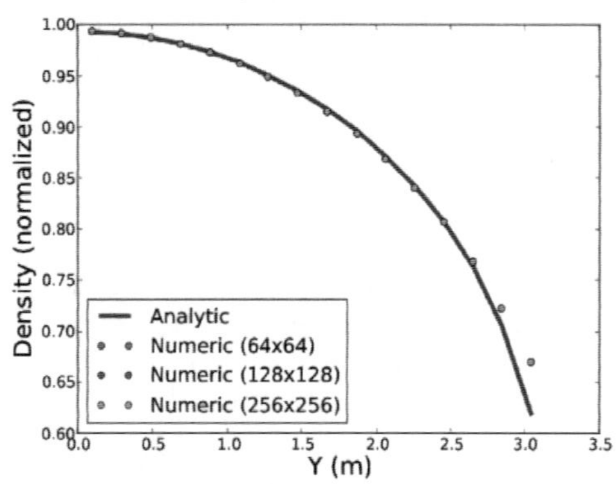

Figure 12. Steady state solution for density in the one dimensional transport model and the analytical solution using the FCI method on straight field lines

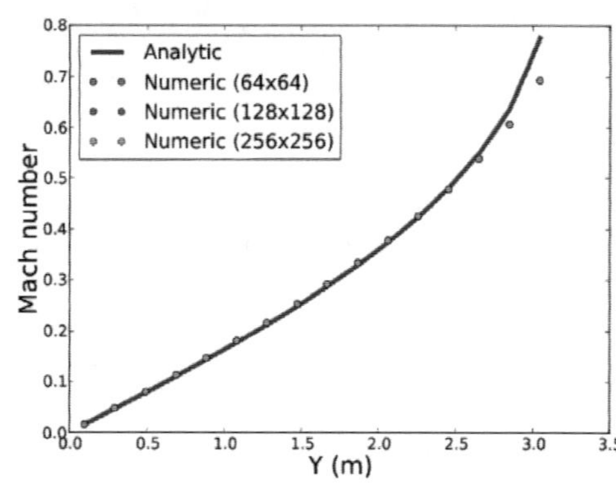

Figure 13. Comparison of the steady state velocity and an analytical solution using the FCI method in a region of straight field lines.

distance to the edge of the computational domain, where the shear causes a shift of 2.35cm in the azimuthal direction between each perpendicular plane (separated by 19.6cm).

Again, the simulations of the one dimensional transport model have reproduced the analytical solution along helical field lines, when considering a sufficiently high resolution (recalling Section 3.2). As these helical field lines require the use of interpolation in the FCI operators, these results indicate that errors due to interpolation are reduced to tolerable levels at sufficient resolution and the FCI method is capable of simulating transport models.

5. Conclusions

Here we have discussed the recent progress in modelling nonaxisymmetric geometries within BOUT++. The Flux Coordinate Independent approach to parallel derivatives has been implemented into BOUT++ and allows for complex geometries to be modelled. A very large aspect ratio classical stellarator test case was implemented and it was determined that the FCI approach

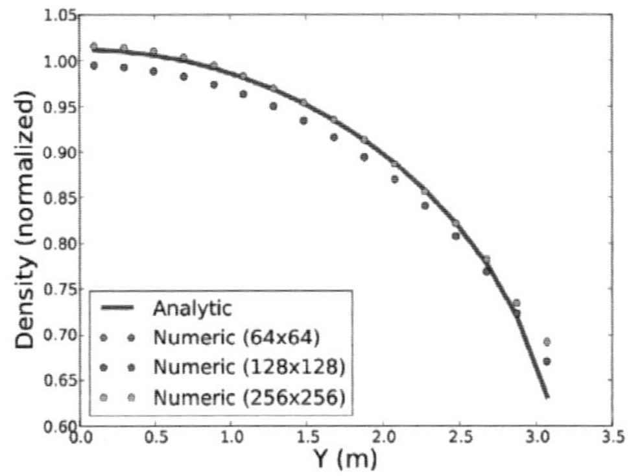

Figure 14. Steady state density and the analytical solution of a one dimensional transport model in a region of helical field lines for three different resolutions. Accuracy is increased with resolution.

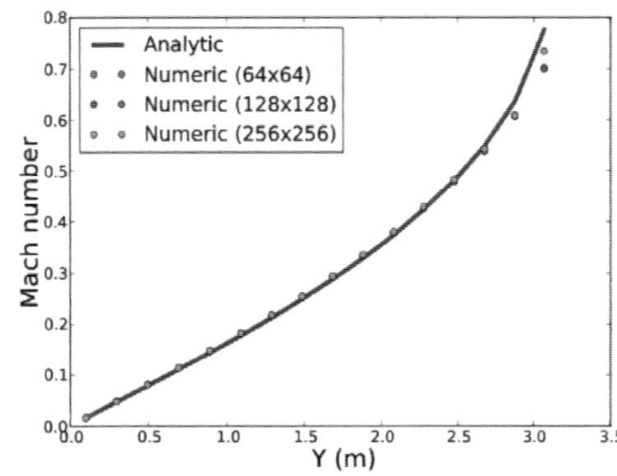

Figure 15. Comparison of the steady state velocity and an analytical solution in a region of helical field lines. Again, the higher resolution cases provide more accurate results.

is correctly evaluating parallel dynamics, which was the most difficult challenge in modelling nonaxisymmetric configurations. A one dimensional transport model was implemented and tested against an analytic solution, where it was determined that BOUT++ can effectively converge to the analytical solution using FCI operators. These results indicate that BOUT++ has the components necessary to model nonaxisymmetric cases.

6. Acknowledgments

This work has been carried out within the framework of the EUROfusion Consortium and has received funding from the Euratom research and training programme 2014-2018 under grant agreement No 633053. The views and opinions expressed herein do not necessarily reflect those of the European Commission.

References

[1] Darwin D-M Ho and Russell M Kulsrud. Neoclassical transport in stellarators. *Physics of Fluids (1958-1988)*, 30(2):442–461, 1987.
[2] G Grieger, W Lotz, P Merkel, J Nührenberg, J Sapper, E Strumberger, H Wobig, R Burhenn, V Erckmann, U Gasparino, et al. Physics optimization of stellarators. *Physics of Fluids B: Plasma Physics (1989-1993)*, 4(7):2081–2091, 1992.
[3] J Nührenberg, W Lotz, P Merkel, C Nührenberg, U Schwenn, E Strumberger, and T Hayashi. Overview on wendelstein 7-x theory. *Fusion Technology*, 27(CONF-941182–), 1995.
[4] T Görler, X Lapillonne, S Brunner, Tilman Dannert, Frank Jenko, Florian Merz, and D Told. The global version of the gyrokinetic turbulence code gene. *Journal of Computational Physics*, 230(18):7053–7071, 2011.
[5] Y Feng, F Sardei, and J Kisslinger. 3d fluid modelling of the edge plasma by means of a monte carlo technique. *Journal of nuclear materials*, 266:812–818, 1999.
[6] Neil McTaggart, Roman Zagórski, Xavier Bonnin, A Runov, Ralf Schneider, T Kaiser, T Rognlien, and M Umansky. 3d edge energy transport in stellarator configurations. *Journal of nuclear materials*, 337:221–226, 2005.
[7] B W Shanahan and B D Dudson. X-point modelling in linear configurations using bout++. *Journal of Physics: Conference Series*, 561(1):012015, 2014.
[8] Brendan Shanahan and Ben Dudson. Blob dynamics in torpex poloidal null configurations. *arXiv preprint arXiv:1605.00963*, 2016.

[9] Jarrod Leddy, Ben Dudson, Michele Romanelli, Brendan Shanahan, and Nick Walkden. A novel flexible field-aligned coordinate system for tokamak edge plasma simulation. *arXiv preprint arXiv:1604.05876*, 2016.

[10] Peter Hill, Brendan Shanahan, and Ben Dudson. Dirichlet boundary conditions for arbitrary-shaped boundaries in stellarator-like magnetic fields for the flux-coordinate independent method. *arXiv preprint arXiv:1608.02416*, 2016.

[11] F Hariri and M Ottaviani. A flux-coordinate independent field-aligned approach to plasma turbulence simulations. *Computer Physics Communications*, 184(11):2419 – 2429, 2013.

[12] Andreas Stegmeir, David Coster, Omar Maj, Klaus Hallatschek, and Karl Lackner. The field line map approach for simulations of magnetically confined plasmas. *Computer Physics Communications*, 198:139–153, 2016.

[13] M. Held, M. Wiesenberger, and A. Stegmeir. Three discontinuous galerkin schemes for the anisotropic heat conduction equation on non-aligned grids. *Computer Physics Communications*, 199:29 – 39, 2016.

[14] P Hill, F Hariri, and M Ottaviani. The effect of magnetic islands on ion temperature gradient turbulence driven transport. *Physics of Plasmas (1994-present)*, 22(4):042308, 2015.

[15] S Günter, Q Yu, J Krüger, and K Lackner. Modelling of heat transport in magnetised plasmas using non-aligned coordinates. *Journal of Computational Physics*, 209(1):354–370, 2005.

[16] J Cosfeld, F Hasenbeck, M Rack, and Y Feng. Analytische lösungen des vereinfachten emc3 models. In *DPG-Frühjahrstagung der Sektion Atome, Moleküle, Quantenoptik und Plasmen (SAMOP)*. DPG, 2016.

Runaway-electron formation and electron slide-away in an ITER post-disruption scenario

A Stahl[1], O Embréus[1], M Landreman[2], G Papp[3] and T Fülöp[1]

[1] Department of Physics, Chalmers University of Technology, Göteborg, Sweden
[2] Institute for Research in Electronics and Applied Physics, University of Maryland, College Park, Maryland 20742, USA
[3] Max Planck Institute for Plasma Physics, Garching, Germany

E-mail: stahla@chalmers.se

Abstract. Mitigation of runaway electrons is one of the outstanding issues for the reliable operation of ITER and other large tokamaks, and accurate estimates for the expected runaway-electron energies and current are needed. Previously, linearized tools (which assume the runaway population to be small) have been used to study the runaway dynamics, but these tools are not valid in the cases of most interest, i.e. when the runaway population becomes substantial. We study runaway-electron formation in a post-disruption ITER plasma using the newly developed non-linear code `NORSE`, and describe a feedback mechanism by which a transition to electron slide-away can be induced at field strengths significantly lower than previously expected. If the electric field is actively imposed using the control system, the entire electron population is quickly converted to runaways in the scenario considered. We find the time until the feedback mechanism sets in to be highly dependent on the details of the mechanisms removing heat from the thermal electron population.

1. Introduction

Runaway electrons pose a severe threat to the safety and reliability of ITER and other high-plasma-current fusion devices [1]. The larger the runaway-electron population, the larger the threat to the integrity of the device. However, if the electron momentum-space distribution function becomes highly non-Maxwellian due to the presence of a high-energy tail of runaways, existing numerical tools employing linearized collision operators are no longer valid. The same is true if the electric field (even momentarily) becomes comparable to the Dreicer field [2].

We recently presented `NORSE` [3] – an efficient solver of the kinetic equation in a homogeneous plasma – which includes the full relativistic non-linear collision operator of Braams & Karney [4, 5]. `NORSE` – which will be discussed in Section 2 – is able to model Dreicer and hot-tail runaway generation in the presence of electric fields of arbitrary strength and synchrotron-radiation reaction: one of the most important energy-loss channels for runaways [6].

Since `NORSE` is able to treat highly distorted distributions, a range of new questions may be addressed. One issue of particular interest is: will non-linear phenomena accelerate or dampen the growth of runaways? Naturally, this is of great importance in view of ITER and other large tokamaks, as it potentially impacts the requirements on the disruption mitigation system (the design of which is currently being finalized) [1, 7]. In addition, the electric field is expected to reach values as high as 80-100 V/m during the current quench in ITER [8], and runaway generation is likely to be strong enough for 60% or more of the plasma current to be converted to runaway current. In Section 3, we use NORSE to model the evolution of the electron population in a typical ITER post-disruption scenario.

If the electric field is strong enough, the net parallel force experienced by electrons due to the electric field and collisions becomes positive in the entire momentum space, leading to a phenomenon known as electron *slide-away*. This is expected to happen when $E > 0.215E_{\rm D} \equiv E_{\rm SA}$, where $E_{\rm D} = ne^3 \ln\Lambda/4\pi\epsilon_0^2 T$ is the so-called Dreicer field [2]; n, T and $-e$ are the electron number density, temperature and charge; $\ln\Lambda$ is the Coulomb logarithm; and ϵ_0 is the vacuum permittivity. The associated surge in runaway current can have a large impact on the potential for material damage, as well as the subsequent evolution of the parallel electric field. The slide-away process cannot be consistently modelled using linear tools such as CODE [9, 10] or LUKE [11], which assume a Maxwellian background plasma and therefore require $E \ll E_{\rm SA}$, as well as that the runaway fraction is small $n_{\rm r}/n \ll 1$.

A strong electric field represents a source of energy that quickly heats up the electron distribution. This heating can induce a transition to the slide-away regime – even under a fixed applied electric field which is initially below the threshold $E < E_{\rm SA}$ – since the collisional friction is lower in a hotter distribution. As a consequence, the Dreicer field is also lower, making the effective normalized field $E/E_{\rm D,eff}$ higher for a given field strength. If the temperature increase is large enough, the slide-away regime is reached, which happens at a field of $E/E_{\rm D,eff} = 0.215$ in the case of a constant applied electric field, i.e. it coincides with the standard slide-away field at the effective temperature $T_{\rm eff}$ [3].

In practice, many processes act to remove heat from the plasma. In a cold post-disruption plasma, line radiation and bremsstrahlung from interactions with partially ionized impurities are important loss channels, as is radial heat transport. Including a heat sink in numerical simulation of such scenarios is therefore desirable, and the sink effectively acts to delay or prevent the transition to slide-away. In this paper, we demonstrate that the evolution of the runaway electron population – including the time to reach slide-away – is highly sensitive to the properties of the applied heat sink, making a detailed investigation of the various loss channels an area of interest for future work.

2. NORSE

We will use the newly developed fully relativistic non-linear tool NORSE [3] to study the dynamics of the electron population. NORSE, which is valid in spatially uniform plasmas, solves the kinetic equation

$$\frac{\partial f}{\partial t} - \frac{e\mathbf{E}}{m_{\rm e}c} \cdot \frac{\partial f}{\partial \mathbf{p}} + \frac{\partial}{\partial \mathbf{p}} \cdot (\mathbf{F}_{\rm s} f) = C_{\rm ee}\{f\} + C_{\rm ei}\{f\} + S, \tag{1}$$

where f is the electron distribution function, t is the time, $m_{\rm e}$ and $\mathbf{p}$ are the electron rest mass and momentum, $\mathbf{E}$ is the electric field, c is the speed of light, $\mathbf{F}_{\rm s}$ is the synchrotron-radiation-reaction force, $C_{\rm ee}$ is the relativistic non-linear electron-electron collision operator, $C_{\rm ei}$ is the electron-ion collision operator, and S represents heat and particle sources or sinks. For a detailed description of the various terms and operators, see Ref. [3]. For the remainder of this paper, we define the electric field such that electrons are accelerated in the positive $p_\|$ direction.

In NORSE, the particle momentum $\mathbf{p}$ is represented in terms of the magnitude of the normalized momentum $p = \gamma v/c$ (where v is the velocity of the particle and γ is the relativistic mass factor) and the cosine of the pitch angle $\xi = p_\|/p$. The kinetic equation is discretized using finite differences in both p and ξ. A linearly implicit time-advancement scheme is used, where the five relativistic Braams-Karney potentials [4] – analogous to the Rosenbluth potentials in the non-relativistic case – are calculated explicitly from the known distribution. These are then used to construct the electron-electron collision operator $C_{\rm ee}$, and the remainder of the kinetic equation is solved implicitly.

For the results presented in this paper, a non-uniform finite-difference grid was used to improve the computational efficiency. In the pitch-angle coordinate, the grid points were chosen with a dense spacing close to $\xi = \pm 1$ – in particular at $\xi = 1$ where the runaway tail forms – but with a sparser grid at intermediate ξ. In the p-direction, a grid mapping with a tanh step in spacing was used in order to produce a grid with dense spacing at low momenta (to accurately resolve the bulk dynamics), and larger spacing in the high-energy tail, where the scale length of variations in f is larger.

The runaway region in NORSE was determined by studying particle trajectories in phase space, neglecting momentum-space diffusion but including self-consistent collisional friction and synchrotron-radiation reaction [3]. The trajectory that terminates at $\xi = 1$ and $p = p_c$ marks the lower boundary of the runaway region, since particles that follow it neither end up in the bulk nor reach arbitrarily high energies. The critical momentum p_c is the momentum at which the balance of forces in the parallel direction ($\xi = 1$) becomes positive (i.e. the lowest momentum at which the accelerating force of the electric field overcomes the collisional and synchrotron-radiation-reaction drag). If the balance of forces is positive for all p at $\xi = 1$, however, all electrons experience a net acceleration, and the population is in the slide-away regime.

2.1. Heat sink

Including a heat sink (HS) in the numerical simulations is of great importance for accurate modelling of the distribution evolution during a disruption. The heat sink used in NORSE to remove heat from the thermal population has the form $S = \partial/\partial\mathbf{p} \cdot (k_\mathrm{h}\mathbf{S}_\mathrm{h} f)$, where $\mathbf{S}_\mathrm{h}(p)$ is an isotropic function of momentum (i.e. $\mathbf{S}_\mathrm{h} \parallel \mathbf{p}$) and $k_\mathrm{h}(t)$ is the magnitude of the source. The terms in the kinetic equation that affect the total energy content are the electric-field and synchrotron-radiation-reaction terms, however; when considering a subset $\mathbf{\Omega}$ of momentum space, collisions can also transfer energy in or out of $\mathbf{\Omega}$, and a corresponding term must be included. The total energy change $\mathrm{d}W/\mathrm{d}t$ in $\mathbf{\Omega}$ can thus be written as

$$\frac{\mathrm{d}W}{\mathrm{d}t} = m_\mathrm{e}c^2 \int_{\mathbf{\Omega}} \mathrm{d}^3 p\,(\gamma - 1)\left(-\frac{e\mathbf{E}}{m_\mathrm{e}c} \cdot \frac{\partial f}{\partial \mathbf{p}} + \frac{\partial}{\partial \mathbf{p}} \cdot (\mathbf{F}_\mathrm{s} f) - C\{f\} + k_\mathrm{h}\frac{\partial}{\partial \mathbf{p}} \cdot (\mathbf{S}_\mathrm{h} f) \right). \tag{2}$$

The magnitude k_h of the sink in each time step can be determined by requiring $\mathrm{d}W/\mathrm{d}t = 0$. In this work, we take $\mathbf{\Omega}$ to represent the thermal bulk of the distribution, which we define as all particles with $v < 4v_{\mathrm{th},0}$, with $v_{\mathrm{th},0} = \sqrt{2T_0/m_\mathrm{e}}$ the thermal speed at the initial temperature T_0.

In this study, the p-component of $\mathbf{S}_\mathrm{h}$ was chosen to have the shape of a Maxwellian at the desired temperature T. In practice, the momentum dependence of the sink will be more complicated and subject to the details of the particular physical processes at work. It will also likely have a limited energy-removal rate, dictated by for instance spatial gradients or impurity content, which could limit its efficiency in maintaining a given temperature. A detailed investigation of the characteristics of the sink is left for future work; the aim of this paper is to highlight the sensitivity of the runaway-electron evolution to the particulars of the sink, and for that purpose we will impose a limit on the energy-removal rate, as will be discussed in the next section.

3. Runaway generation in an ITER disruption

3.1. Post-disruption scenario

In this section, we use NORSE to model the evolution of the electron distribution during a typical ITER disruption. The electric field evolution (which is shown in Fig. 1a) and other parameters are taken from the ITER inductive scenario no. 2 in Ref. [8], but the temperature evolution has been simplified to facilitate the numerical calculation. We assume the electron population to be completely thermalized and use the final temperature $T = T_0 = 10$ eV throughout our

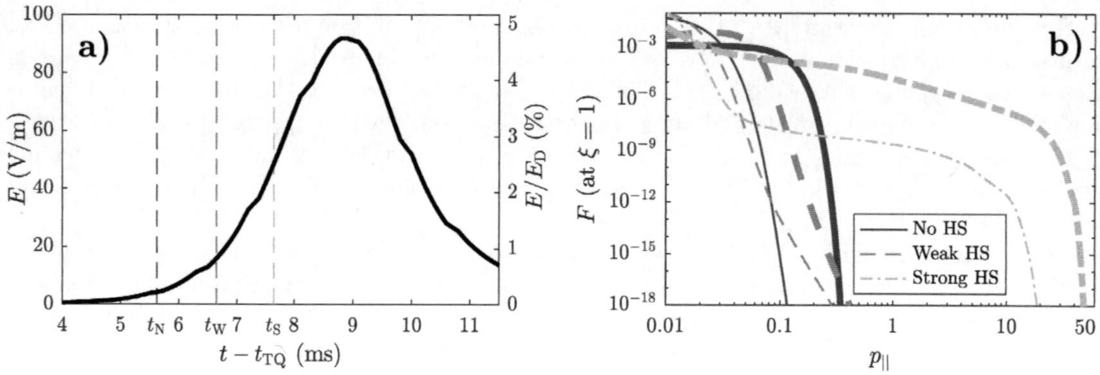

Figure 1. a) Electric field in V/m (left vertical axis) and normalized to the Dreicer field E_D at the temperature T_0 and density n_0 (right vertical axis), as a function of time after the thermal quench. b) Tail of the parallel electron distribution. Thin lines show f at t_N (no HS), t_W (weak HS) and t_S (strong HS), and thick lines show f immediately before the transition to slide-away.

simulation, together with the density $n = n_0 = 7.1 \cdot 10^{19}$ m^{-3}. This is likely to underestimate the runaway-electron generation in the early phase, in which the temperature is still dropping, however the chosen set-up is sufficient for our purposes. We use the magnetic field on axis ($B = 5.3$ T) and $Z_{\rm eff} = 1$. The initial current density in the scenario is $j_0 = 0.62\,\mathrm{MA/m^2}$, however our calculations start from a Maxwellian distribution and make no attempt to maintain the experimental current evolution explicitly.

To highlight the importance of the temperature evolution of the bulk, in Section 3.2 we will consider three scenarios: *no heat sink* (subscript N), *weak heat sink* (W) and *strong heat sink* (S). In the no-heat-sink scenario, all the energy supplied by the electric field will remain in the simulation, leading to rapid bulk heating; with the strong heat sink, a bulk temperature of $T_0 = 10$ eV will be enforced in accordance with Eq. 2, i.e. any excess heat in the bulk region will be removed using a heat sink. In the intermediate case of a weak heat sink, the energy-removal rate of the heat sink will be restricted to 0.5 MW/m^3. This particlar value has been chosen at will, but is meant to represent some inherent limitation in the physical processes responsible for the energy loss. In both the weak and strong cases, the heat sink will affect only the thermal population, allowing the supra-thermal tail to gain energy from the electric field. Physically, this corresponds to processes not included in the simulation (such as radial transport or radiative losses) which primarily affect the thermal population.

NORSE simulations of the evolution of the electron distribution function in the presence of the electric field in Fig. 1a were performed for the three different scenarios. The simulations were aborted when the runaway population reached $n_r/n = 1$; i.e. a transition to the slide-away regime was observed. The simulation results can however only be considered characteristic of a natural ITER disruption for current densities comparable to, or somewhat larger than, the initial value j_0, since after that point the strong response of the inductive electric field to the increased local current would invalidate the E-field evolution used. We will therefore mark the time where the current density reaches $j/j_0 > 5$ in all plots, and denote it with t_N, t_W and t_S, respectively, for the no-sink, weak-sink and strong-sink scenarios. The distribution evolution at later times can only be considered accessible in scenarios where the loop voltage is actively sustained using the control system. Nevertheless, this regime will turn out to be of interest, since a non-linear feedback mechanism leading to a rapid transition to slide-away is observed.

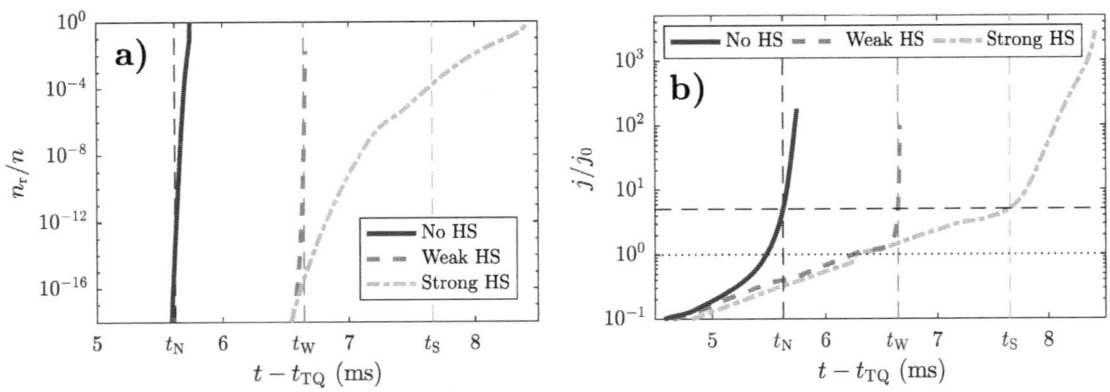

Figure 2. a) Runaway fraction and b) current density normalized to its initial value j_0, as a function of time after the thermal quench in the different heat-sink scenarios. The times t_N, t_W and t_S (vertical thin dashed lines), mark the time where the current density reaches $j/j_0 > 5$ for the no-heat-sink, weak-heat-sink and strong-heat-sink scenarios, respectively.

3.2. Evolution of the runaway-electron population

In Fig. 1b, the tails of the distributions in the parallel direction are shown at t_N, t_W and t_S (thin lines), as well as at the final times (thick lines) in each scenario (just before the transition to slide-away is reached). In the figure, the distribution is normalized such that $F = f/f_M(t{=}0, p{=}0)$, where f_M is a Maxwell-Jüttner distribution, so that F initially takes the value unity at $p = 0$. The maximum achieved particle energies are highly dependent on whether a heat sink was applied or not; in the no-heat-sink and weak-heat-sink scenarios, the particles did not have time to reach relativistic energies, whereas in the strong heat sink case, $p \approx 19$ and $p \approx 44$ (corresponding to energies of roughly 9 and 22 MeV), were obtained at t_S and just before reaching slide-away, respectively. The reason for this is that, as we shall see, the current density growth and subsequent transition to slide-away in the latter case occur at much later times, and the runaways have time to gain more energy.

Figure 2a shows the evolution of the runaway fraction during the course of the simulation. In the no-heat-sink case, the runaway fraction increases sharply, but as shown in Fig. 1b, the runaways are all at low energy. The transition to slide-away happens already at $t=$ 5.7 ms; early on in the electric-field evolution (cf. Fig. 1). In the two scenarios employing a heat sink, the growth in runaway fraction occurs later, but in the weak-heat-sink case the growth rate is comparable to the case when a sink is absent once the process is initiated. In this case, the transition to slide-away happens at $t = 6.7$ ms. With the strong heat sink, the runaway population grows steadily, eventually dominating the entire distribution at $t = 8.4$ ms, however in this case the transition is gradual, rather than rapid. In all three scenarios, including the one with an ideal strong heat sink, the slide-away regime is thus reached even before the electric field (calculated assuming a linear treatment) has reached its peak.

Figure 2b shows the evolution of the current density. It indicates that the rapid increase in the runaway fraction is correlated with a similar increase in the current density, although in the no-sink case, the growth rate is somewhat smaller. Again, the growth in the strong-heat-sink case is gradual, rather than explosive. Note that in the no and weak heat-sink cases, the runaway fraction is still negligibly small at the start of the rapid transition to slide-away. The transition is thus not a non-linear phenomenon triggered by the size of the runaway population; it starts in a regime where linearized tools are normally expected to be valid, and before the current density becomes significantly larger than its initial value.

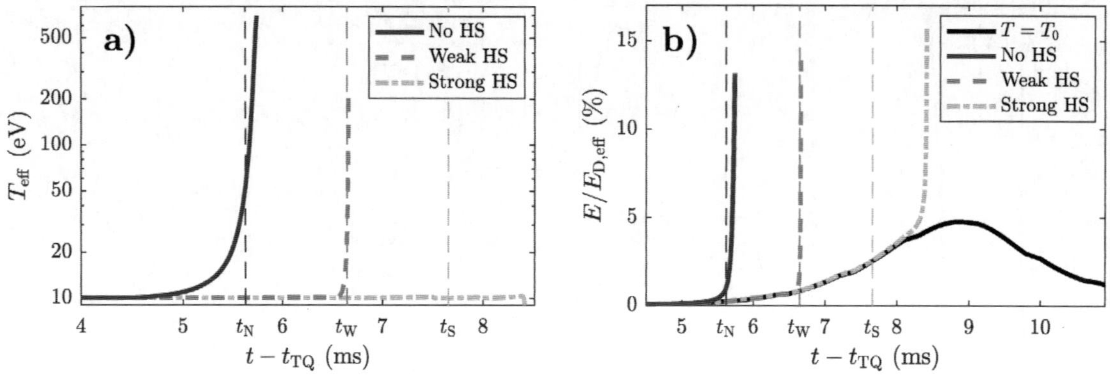

Figure 3. a) Effective temperature of the bulk population and b) corresponding effective normalized electric-field strength (also taking changes to the bulk density into account) as functions of time, in the different scenarios. The black solid line in panel b) corresponds to the normalized field with $T = T_0$ and $n = n_0$ (cf. Fig. 1).

The explanation can be found by examining the thermal population. By comparing the energy moment of the bulk of the distribution ($W_{\boldsymbol{\Omega}}$) with that of a relativistic Maxwellian ($W_{\rm M}(T)$),

$$W_{\boldsymbol{\Omega}} = m_{\rm e}c^2 \int_{\boldsymbol{\Omega}} d^3p\,(\gamma - 1)f = W_{\rm M} = \frac{m_{\rm e}c^2 n}{\Theta_{\rm eff}K_2(1/\Theta_{\rm eff})} \int_0^{p_{\rm max,\boldsymbol{\Omega}}} {\rm d}p\, p^2(\gamma - 1)\exp\left(-\frac{\gamma}{\Theta_{\rm eff}}\right), \quad (3)$$

an effective temperature $T_{\rm eff}$ for a given distribution f can be determined by solving for $\Theta_{\rm eff} = T_{\rm eff}/m_{\rm e}c^2$. In the above equation, $p_{\rm max,\boldsymbol{\Omega}}$ is the upper boundary in p of the bulk region in momentum space, and $K_2(x)$ is the modified Bessel function of the second kind (and order two). The effective temperature $T_{\rm eff}$ is plotted in Fig. 3a as a function of time for all three scenarios. In the no-heat sink case, it increases by roughly two orders of magnitude during the simulation, i.e. when energy is not actively removed from the system, the Ohmic heating is sufficient to heat the plasma to a temperature of about 700 eV before the onset of slide-away, or 55 eV if the electric field is not artificially sustained. A similar (albeit weaker) tendency is seen in the weak-heat-sink case, where the temperature increases to $T_{\rm eff} \approx 210\,{\rm eV}$ in the phase leading up to the slide-away transition, or 25 eV in a non-driven case. This heating is a consequence of the imposed limited maximum energy-removal rate of the heat sink in the weak case, since the temperature is efficiently kept constant in the beginning of the simulation, where $({\rm d}W/{\rm d}t)_{\rm HS} < 0.5\,{\rm MW/m^3}$. The strong heat sink manages to keep $T_{\rm eff} - T_0$ to within a few tenths of an eV during the entire simulation, corresponding to a source with unlimited (or at least higher than required) maximal energy-removal rate.

The significance of the observed bulk heating is its influence on the Dreicer field $E_{\rm D}$, which (apart from the weak dependence on $\ln\Lambda$) is inversely proportional to T. For a given electric-field strength, the normalized field $E/E_{\rm D}$ thus increases as the bulk heats up. The effective normalized electric field is shown in Fig. 3b, indicating that the rapid growth in the runaway fraction in Fig. 2a is correlated with a sudden increase in the normalized electric field in the no-heat-sink and weak-heat-sink cases. In the strong-heat-sink case, the increase in normalized field is not caused by the temperature, which is kept constant during the entire simulation, but by the decrease in the bulk density as the runaway population becomes substantial. As can be seen from the yellow dash-dotted line in the figure, this has a similar effect as a temperature increase, since $E_{\rm D} \sim n_{\rm bulk}$. The effective $E/E_{\rm D}$ starts to deviate from the baseline value (black solid line)

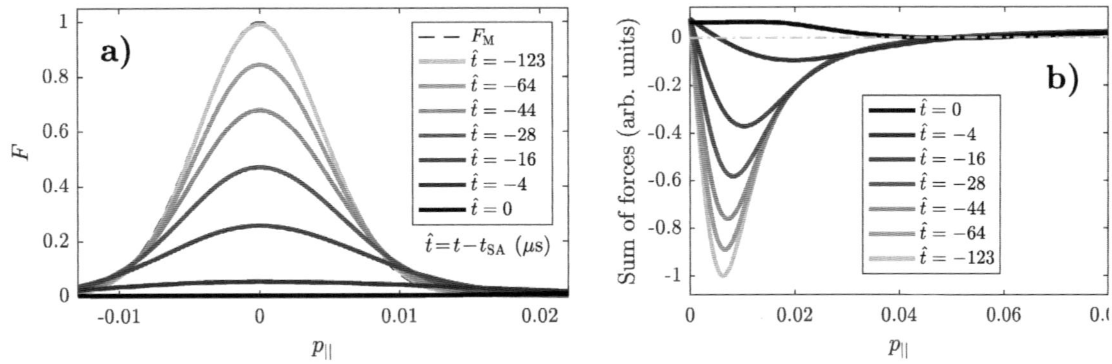

Figure 4. a) Evolution of the bulk of the electron distribution and b) balance of forces, in the direction parallel to the electric field, just before the transition to slide-away in the weak-heat-sink scenario. $\hat{t}$ is the time relative to the transition to slide-away in μs.

already when $n_{\rm bulk}/n_0 \approx 0.97$. The feedback process is thus initiated when the runaway fraction is just 3%; a regime where linear tools are expected to be valid.

The two effects of increasing $T_{\rm eff}$ and decreasing $n_{\rm bulk}$ lead to a positive feedback mechanism which is responsible for the rapid growth in runaway fraction and current density seen in the no-heat-sink and weak-heat-sink cases. Once the bulk temperature has increased enough (or a high enough runaway tail is produced, as seen in the strong-heat-sink case), the normalized electric field becomes strong enough to cause a depletion of the bulk through primary runaway generation. The reduced bulk density in turn leads to a more efficient heating of the remaining bulk particles. Both of these effects contribute to a reduction in the Drecier field $E_{\rm D}$, and a corresponding reduced collisional friction on the bulk electrons, which makes runaway acceleration easier. This further increases the rate of bulk depletion, and so on. Eventually the friction becomes low enough that the parallel balance of forces becomes positive everywhere, marking the transition to the slide-away regime. At this point, the bulk of the distribution can no longer be well described by a Maxwellian and the positive feedback mechanism makes the transition possible even though $E/E_{\rm D,eff}<0.215$ (in this case at around $E/E_{\rm D,eff}\approx 0.15$).

The bulk depletion and associated change in the parallel force balance is shown in Fig. 4a and Fig. 4b, respectively, in the phase leading up to the transition to slide-away in the weak-heat-sink case. Initially, the force balance is positive in the tail – i.e. particles there experience a net acceleration – while it is negative in the bulk, meaning particles are slowed down by collisions. As the feedback process starts, the minimum in the sum of forces becomes gradually less pronounced, in tandem with the depletion of the bulk population, up until the point where the sum of forces becomes positive everywhere, and the slide-away regime is reached. This highly non-linear process cannot be accurately captured by a linear model.

4. Discussion and conclusions

In this paper we have examined the evolution of the electron distribution and runaway generation in an ITER-like post-disruption scenario where the electric fields reach values as high as 90 V/m. With the help of the newly developed tool `NORSE`, which is a relativistic non-linear solver for the electron momentum-space distribution function, we have shown that the slide-away regime, i.e. a net parallel acceleration of electrons in all of momentum space, is reached in this scenario, provided the electric field evolution used is artificially enforced by the control system. In the stage leading up to the transition, a positive feedback mechanism sets in by means of which the bulk quickly gets depleted by primary runaway generation which reduces the friction on the thermal population, leading to further bulk depletion, until the point where the slide-away is reached. This process can be initiated at significantly weaker fields than the slide-away field E_{SA} expected from linear theory.

The time to a transition to slide-away is highly dependent on the ability of loss processes to remove heat from the thermal electron population, but even with an ideal sink (the strong-heat-sink case in Section 3), complete runaway generation was seen 8.4 ms after the thermal quench. These results were obtained without taking avalanche or hot-tail runaway generation into account, which would only lead to more prominent runaway growth.

Also in the case of a disruption where the electric field is not artificially sustained, strong bulk heating leads to a rapid growth in the runaway fraction because of the increase in the normalized electric field E/E_{D}, but the current density becomes large enough to significantly affect the electric field evolution (supressing the growth of E) before the slide-away regime is reached. This is observed in the absence of a heat sink, as well as with a heat sink with a limited maximum energy-removal rate (in this case 0.5 MW/m^3). If the efficiency of the heat sink is not limited, the runaway fraction grows more slowly and the runaways have time to reach significantly higher energies before the electric field becomes affected by the growing current density. The severity of disruptions in ITER could thus be greatly affected by the properties of the heat sinks present in the plasma.

The feedback mechanism described in this paper has important consequences for the understanding of runaway-electron dynamics. With the entire electron population experiencing a net accelerating force at much weaker electric fields than previously expected, very large runaway-electron current generation is likely. This would impact the subsequent electric-field evolution, leading to a reduction in field strength and duration which could occur at realatively early times if the heat sink has a limited energy removal rate. Therefore, it is difficult to determine the magnitude of the effect on the current evolution and post-quench dynamics without a self-consistent calculation of the electron distribution and the electric field. Nevertheless, this paper shows that feedback effects play an important role in post-disruption runaway dynamics, and that the details of the heat-loss channels may have a big impact on what strength and duration of electric field can be tolerated before the positive feedback, and possible subsequent transition to slide-away, is induced.

Acknowledgments

This work was supported by the Swedish Research Council (Dnr. 2014-5510), the European Research Council (ERC-2014-CoG grant 647121), and the Knut and Alice Wallenberg Foundation. M.L. was supported by the U.S. Department of Energy, Office of Science, Office of Fusion Energy Science, under Award Numbers DE-FG02-93ER54197 and DE-FC02-08ER54964.

References

[1] Hollmann E M, Aleynikov P B, Fülöp T, Humphreys D A, Izzo V A, Lehnen M, Lukash V E, Papp G, Pautasso G, Saint-Laurent F and Snipes J A 2015 *Physics of Plasmas* **22** 021802
[2] Dreicer H 1959 *Physical Review* **115**(2) 238–249
[3] Stahl A, Landreman M, Embréus O and Fülöp T 2016 *Submitted to Computer Physics Communications* URL `http://arxiv.org/abs/1608.02742`
[4] Braams B J and Karney C F F 1987 *Physical Review Letters* **59** 1817–1820
[5] Braams B J and Karney C F F 1989 *Physics of Fluids B: Plasma Physics* **1** 1355–1368
[6] Stahl A, Hirvijoki E, Decker J, Embréus O and Fülöp T 2015 *Physical Review Letters* **114** 115002
[7] Boozer A H 2015 *Physics of Plasmas* **22** 032504
[8] Smith H, Helander P, Eriksson L G, Anderson D, Lisak M and Andersson F 2006 *Physics of Plasmas* **13** 102502
[9] Landreman M, Stahl A and Fülöp T 2014 *Computer Physics Communications* **185** 847 – 855 ISSN 0010-4655
[10] Stahl A, Embréus O, Papp G, Landreman M and Fülöp T 2016 *Nuclear Fusion* **56** 112009
[11] Decker J and Peysson Y 2004 DKE: A fast numerical solver for the 3D drift kinetic equation Tech. Rep. EUR-CEA-FC-1736 Euratom-CEA

Fluid and gyrokinetic modelling of particle transport in plasmas with hollow density profiles

D Tegnered, M Oberparleiter, H Nordman and P Strand

Department of Earth and Space Sciences, Chalmers University of Technology, SE-412 96 Gothenburg, Sweden

E-mail: tegnered@chalmers.se

Abstract. Hollow density profiles occur in connection with pellet fuelling and L to H transitions. A positive density gradient could potentially stabilize the turbulence or change the relation between convective and diffusive fluxes, thereby reducing the turbulent transport of particles towards the center, making the fuelling scheme inefficient. In the present work, the particle transport driven by ITG/TE mode turbulence in regions of hollow density profiles is studied by fluid as well as gyrokinetic simulations. The fluid model used, an extended version of the Weiland transport model, Extended Drift Wave Model (EDWM), incorporates an arbitrary number of ion species in a multi-fluid description, and an extended wavelength spectrum. The fluid model, which is fast and hence suitable for use in predictive simulations, is compared to gyrokinetic simulations using the code GENE. Typical tokamak parameters are used based on the Cyclone Base Case. Parameter scans in key plasma parameters like plasma β, R/L_T, and magnetic shear are investigated. It is found that β in particular has a stabilizing effect in the negative R/L_n region, both nonlinear GENE and EDWM show a decrease in inward flux for negative R/L_n and a change of direction from inward to outward for positive R/L_n. This might have serious consequences for pellet fuelling of high β plasmas.

1. Introduction

The characteristics of particle transport in regions of hollow density profiles is an important issue for fusion plasmas. For example, reactor grade plasmas will likely be fuelled by pellet injection that will transiently perturb the density and temperature profiles, making the density profile hollow. A similar effect on the density profile may occur in connection with the L-mode to H-mode transition. Hence, it is important to study the microstability and transport properties in regions with a hollow density profile. In a microstability study of pellet fuelled discharges in MAST [1] using the local gyrokinetic code GS2 [2] in linear mode, a stabilization of all modes was found in the negative R/L_n region as a result of the reduction in η_i and η_e, as well as a high sensitivity to the rapid change in plasma β and collisionality that was caused by the pellet ablation. In [3] a stabilizing effect was also found for pellet fuelled discharges at JET. The linear growth rates as well as the nonlinear heat fluxes were shown to be reduced in the positive gradient region. The quasilinear gyrokinetic code QuaLiKiz [4] was used in [5] to study the microturbulence during the L to H transition, it was shown that the TE mode was stabilized and that the particle flux was highly sensitive to the sign of R/L_n as well as the collisionality and R/L_T.

In order to study the effects the positive density gradient ($R/L_n < 0$ where $R/L_n = -R\nabla n/n$)

Table 1: Typical CBC parameters

Parameter	Value
Collisions	No (unless noted)
β	No (unless noted)
q_0	1.4
$\hat{s}$	0.8
B_0	3.1 T
r/R	0.18
R	1.65 m
$T_e = T_i$	2.85 keV
$n_e = n_i$	$3.5 \times 10^{19}\,\mathrm{m}^{-3}$
$R/L_T = R/L_{T_i} = R/L_{T_e}$	6.96

and parametric scans have on the transport the gyrokinetic code GENE [6] is used in local mode to study the transport due to Ion Temperature Gradient (ITG)/Trapped Electron (TE) mode [7, 8, 9, 10, 11, 12, 13] turbulence. Parameter scans around those of the Cyclone Base Case (CBC) [14] are used. The CBC parameters are shown in Table 1. The linear and nonlinear GENE results are compared to results from a fluid model, an extended version of the Weiland [15] transport model, the Extended Drift Wave Model (EDWM) [16].

2. GENE and EDWM simulations setup

GENE solves the nonlinear gyrokinetic Vlasov equations together with Maxwell's equations in order to find the distribution functions of the species, $f\left(\mathbf{R}, v_\parallel, \mu, t\right)$, the electrostatic potential, $\phi(\mathbf{x}, t)$ and the parallel components of the magnetic vector potential and magnetic field, $A_\parallel(\mathbf{x}, t)$ and $B_\parallel(\mathbf{x}, t)$. It is a Eulerian δf-type code where the coordinate system is aligned to the background magnetic field with x as the radial coordinate, y as the binormal coordinate, and z as the parallel coordinate. Both linear and nonlinear simulations are performed with kinetic ions and electrons. In the collisional cases, the collisions are modelled using a linearised Landau-Boltzmann collision operator [17]. The pressure gradient, as used in the calculation of the curvature and ∇B drift, is set to be consistent with the density and temperature gradients and the plasma β. In this work, a circular geometry model is used. For the nonlinear GENE simulations, a simulation domain in the perpendicular plane of $[L_x, L_y] = [126, 126]$ is used, with a resolution of $[n_x, n_y] = [96, 48]$. In the parallel direction 32 grid points are used, and in the parallel velocity direction 64 grid points, and 16 magnetic moments. The simulations are typically run up to a simulation time of $t = 300R/c_s$ where R is the major radius and $c_s = \sqrt{T_e/m_i}$. The resolution and simulation domain are checked through convergence tests.

EDWM is an extended version of the Weiland [18, 15] model which incorporates an arbitrary number of ion species in a multi-fluid description and an extended wavelength spectrum. Here, the wavelengths are fixed at $k_y\rho_s = 0.2$, for the comparison to nonlinear GENE since it is the wavelength corresponding to the largest fluxes, or $k_y\rho_s = 0.3$ for the linear comparison since it is the wavelength corresponding to the largest ITG growth rates. ρ_s is the sound Larmor radius defined as $\rho_s = c_s/(eB/m_s c)$. In the fluid model, the perturbations in density, parallel velocity, and temperature for each species is described by the continuity, parallel momentum and energy equations, including FLR-effects for the main ions. The densities of the main ions, trapped, and free electrons are coupled through the quasineutrality condition. The free electrons are assumed

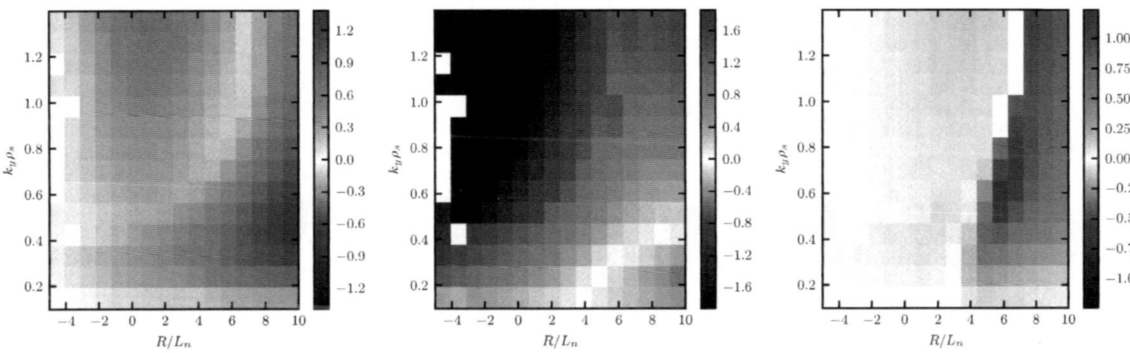

(a) Linear growth rate, γ (b) Linear real frequency, ω_r (c) Normalized flux, $\Gamma/\delta n^2$

Figure 1: GENE eigenvalues as a function of $k_y\rho_s$ and $R/L_{n_e} = R/L_{n_i}$ for the mixed case where $R/L_{T_e} = R/L_{T_i} = 6.96$.

to be Boltzmann distributed in the electrostatic case. A weakly ballooning eigenfunction is assumed whereby the parallel wavenumber $k_\parallel$, the perpendicular wavenumber $k_\perp$ and the magnetic drift ω_{Dj} can be replaced by averages over the weakly ballooning eigenfunction. The particle and heat fluxes are then calculated assuming linear relations between the field quantities combined with a modified mixing length estimate [15].

3. Linear results

The computational parameters used in the linear GENE simulations are a resolution of 48×12 in the parallel and normal direction with 32 grid points in the parallel velocity direction and 24 magnetic moments. An initial value solver is used. Three main cases are simulated linearly: the mixed case with $R/L_{T_e} = R/L_{T_i} = 6.96$, a case where the ITG mode is primarily unstable with $R/L_{T_e} = 0, R/L_{T_i} = 6.96$ and a case where the TE mode is primarily unstable with $R/L_{T_e} = 6.96, R/L_{T_i} = 0.0$.

In Figure 1 the linear growth rates, real frequencies and normalized particle fluxes $\Gamma/\delta n^2$ are shown for the mixed case as a function of the normalized density gradient scale length R/L_n and the normalized poloidal wave number $k_y\rho_s$. In the mixed case, the ITG mode is dominating roughly in the area $k_y\rho_s < 0.5$ and $R/L_n < 4$ and the area $k_y\rho_s > 0.5$ and $R/L_n > 6$ as indicated by the positive values of the real frequencies in Figure 1b. Here a positive ω_r means that the mode is drifting in ion diamagnetic direction, indicating an ITG mode. The TE mode is dominant elsewhere, as indicated by the negative real frequency. In Figure 1c the direction of the particle flux is indicated at different wave numbers and density gradients in the mixed case. Positive values correspond to an outward particle flux while negative correspond to an inward one. The gradient of zero particle flux, indicating the background peaking factor, increases with wave number from $PF \approx 3$ at $k_y\rho_s = 0.2$ to 6 at $k_y\rho_s = 1.2$.

Moving to the pure ITG case in Figure 2 the eigenvalues corresponding to the ITG mode are similar, but the spectra lacks the TE mode that was previously visible for lower R/L_n and higher wave numbers. There is no clear region with inward particle flux as in the mixed case.

Finally, in Figure 3 the eigenvalue spectra of the TE case is examined. Here, the TE mode dominates everywhere and is is stabilized at large negative R/L_n. There is a region of inward transport at large negative R/L_n which however coincides with low growth rates since the TE mode is stabilized in the negative R/L_n region.

In Figure 4 the eigenvalues of GENE and EDWM are compared in a scan in R/L_n. In the

Journal of Physics: Conference Series **775** (2016) 012014 doi:10.1088/1742-6596/775/1/012014

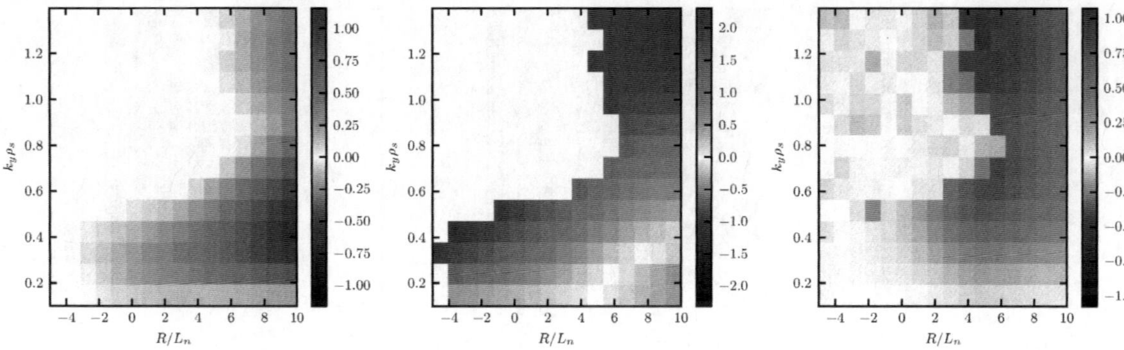

(a) Linear growth rate, γ (b) Linear real frequency, ω_r (c) Normalized flux, $\Gamma/\delta n^2$

Figure 2: GENE eigenvalues as a function of $k_y\rho_s$ and $R/L_{n_e} = R/L_{n_i}$ for the ITG case where $R/L_{T_e} = 0.0$, $R/L_{T_i} = 6.96$.

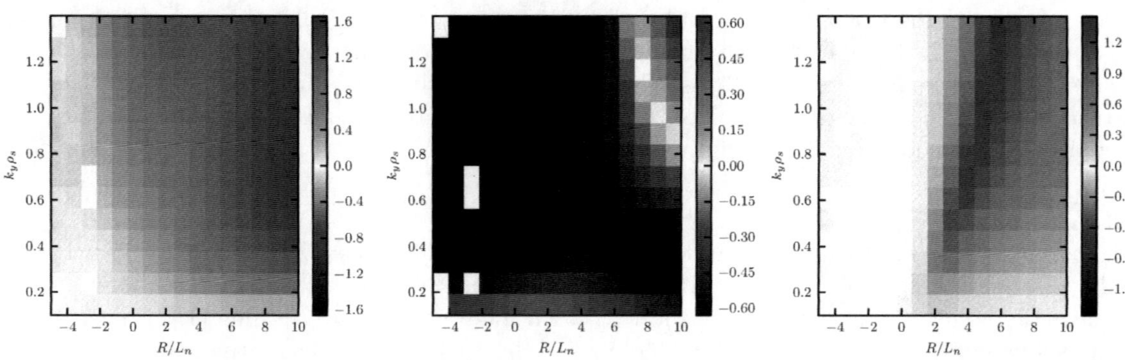

(a) Linear growth rate, γ (b) Linear real frequency, ω_r (c) Normalized flux, $\Gamma/\delta n^2$

Figure 3: GENE eigenvalues as a function of $k_y\rho_s$ and $R/L_{n_e} = R/L_{n_i}$ for the TE case where $R/L_{T_e} = 6.96$, $R/L_{T_i} = 0.0$.

mixed and ITG cases (Figures 4a and 4b) which are both ITG dominated at these wave numbers, the real frequencies match well while the growth rates are significantly reduced in GENE for lower and negative R/L_n. In the TE dominated case, Figure 4c, the growth rates are more similar while the gyrokinetic real frequencies are closer to 0. As an indication of the direction of the gyrokinetic and fluid particle fluxes, $\Gamma/(q_i + q_e)$, where q_i and q_e are the ion and electron heat fluxes, is also shown in Figure 4. Vertical lines indicate the lowest R/L_n where the particle flux changes sign. This matches well between the fluid and gyrokinetic models in the mixed case while in the TE case the R/L_n of zero particle flux is lower for the gyrokinetic model. No negative particle flux is found at this wave number for the pure ITG case.

Next, the dependence on plasma β is explored in Figure 5. β effects in GENE are included as magnetic field fluctuations parallel and perpendicular to the background field, however, since a concentric circular geometry model is used, there is no Shafranov shift. For negative R/L_n β is stabilizing in both models. The MHD ballooning limit can be estimated by $0.6\hat{s}/\left[q_0^2(2R/L_n + R/L_{T_i} + R/L_{T_e})\right]$ [19], so $\beta_{\mathrm{crit,MHD}} = 1.8\%$ for $R/L_n = 0$ and $\beta_{\mathrm{crit,MHD}} = 6.2\%$ at $R/L_n = -5$. Thus, for sufficiently large negative R/L_n, the Kinetic Ballooning Mode

Journal of Physics: Conference Series **775** (2016) 012014 doi:10.1088/1742-6596/775/1/012014

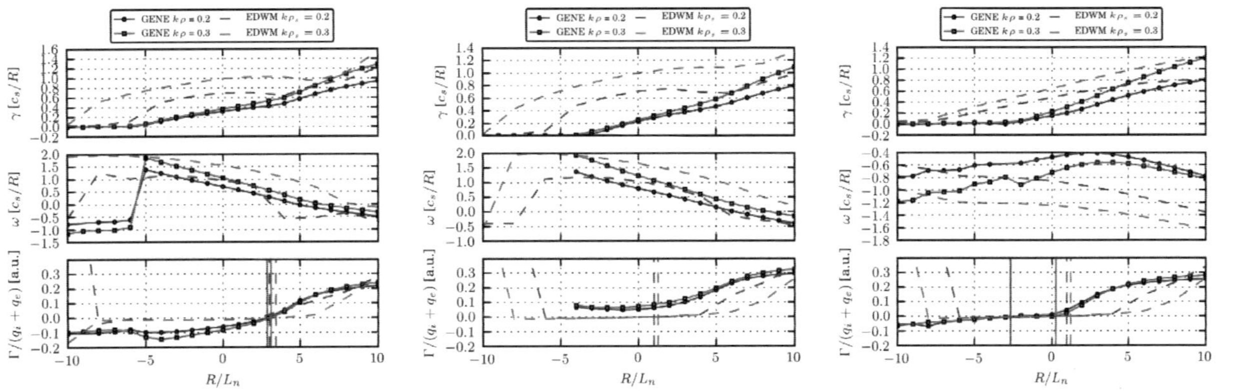

(a) The mixed ITG/TE case ($R/L_{T_e} = R/L_{T_i} = 6.96$)
(b) The pure ITG case ($R/L_{T_e} = 0$, $R/L_{T_i} = 6.96$)
(c) The pure TE case ($R/L_{T_e} = 6.96$, $R/L_{T_i} = 0$)

Figure 4: Comparison of linear gyrokinetic and fluid eigenvalues in scan over R/L_n. Normalized particle flux also shown. Vertical lines indicate value of R/L_n where the particle flux changes sign.

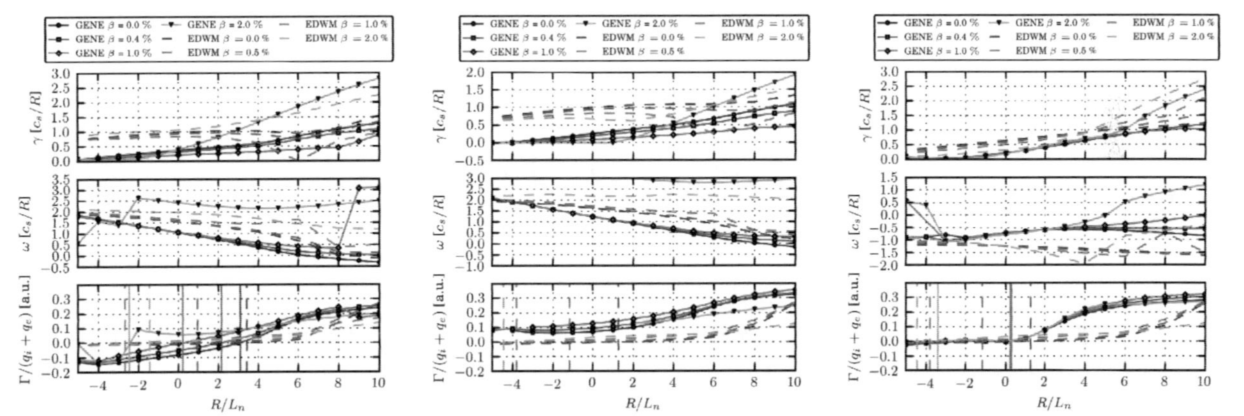

(a) The mixed ITG/TE case ($R/L_{T_e} = R/L_{T_i} = 6.96$)
(b) The pure ITG case ($R/L_{T_e} = 0$, $R/L_{T_i} = 6.96$)
(c) The pure TE case ($R/L_{T_e} = 6.96$, $R/L_{T_i} = 0$)

Figure 5: Linear gyrokinetic and fluid eigenvalues at $k_y\rho_s = 0.3$ in scan over R/L_n and plasma β. Normalized particle flux also shown. Vertical lines indicate value of R/L_n where the particle flux changes sign.

(KBM) will not be excited and the ITG mode will be dominant for which β has a stabilizing effect, which is consistantly found in gyrokinetic simulations [20]. Both models also show a decreasing peaking factor with increasing β in the ITG dominated cases in Figures 5a and 5b, consistent with earlier gyrokinetic findings [21].

In a scaling of collisionality in GENE without changing the temperature or density, it was found that higher collisionality did not stabilize the ITG or TE mode turbulence significantly in the negative R/L_n region. A reduction in the (still positive) background peaking factor was found, which is a known result [21]. A reduction in growth rates with higher magnetic shear in the positive and negative gradient region was found for the ITG and TE dominated cases, along with a reduction in the still positive background peaking factor.

Journal of Physics: Conference Series **775** (2016) 012014 doi:10.1088/1742-6596/775/1/012014

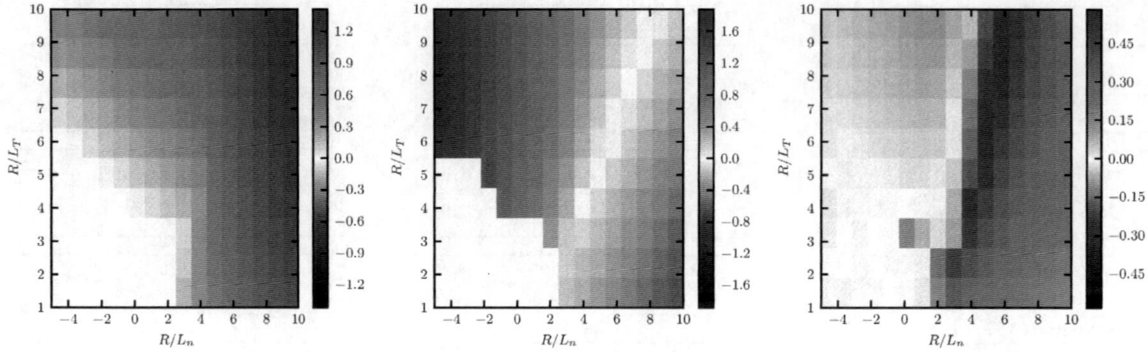

(a) Linear growth rate, γ (b) Linear real frequency, ω_r (c) Normalized flux, $\Gamma/\delta n^2$

Figure 6: GENE eigenvalues as a function of R/L_n and $R/L_{T_i} = R/L_{T_e}$ at $k\rho = 0.3$

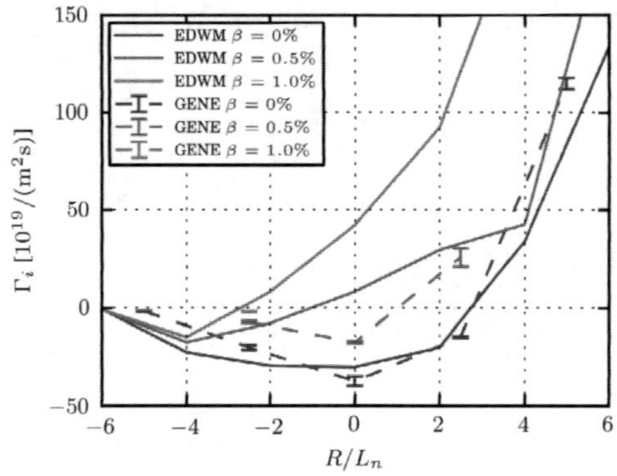

Figure 7: Nonlinear gyrokinetic particle flux as a function of R/L_n and plasma β, compared to EDWM results.

In order to study the turbulence closer to marginal stability, a scan in $R/L_T = R/L_{T_i} = R/L_{T_e}$ and R/L_n is shown in Figure 6. In the negative R/L_n region the ITG mode turbulence will eventually be stabilized with decreasing R/L_T while for higher R/L_n the TE mode turbulence will instead be excited. As shown in Figure 6c, this leads to a lower background peaking factor with the decrease in R/L_T.

4. Nonlinear results

In order to find the saturation level of the turbulence, nonlinear GENE simulations are needed. Nonlinear simulations are performed for the CBC case ($R/L_{T_e} = R/L_{T_i} = 6.96$) and compared to the EDWM results at $k_y\rho_s = 0.2$ for different values of plasma β. The particle fluxes are shown in Figure 7. The fluid and gyrokinetic particle fluxes agree reasonable well at $\beta = 0\%$ and $\beta = 0.5\%$. For $\beta = 0\%$ the particle flux is negative around $R/L_n = 0$ which means that the particles in a pellet ablation peak would travel inwards. However, at higher β the transport decreases in the negative R/L_n region while it changes sign to outward in the positive region. In this parameter regime the pellet fuelling scheme would be less efficient.

Journal of Physics: Conference Series **775** (2016) 012014 doi:10.1088/1742-6596/775/1/012014

5. Conclusions

In this paper the characteristics of particle transport driven by ITG/TE mode turbulence in regions of hollow density profiles were studied by fluid as well as gyrokinetic simulations, using the EDWM and GENE codes in parameter scans around those of the Cyclone Base Case. In the linear analysis, three main cases were studied, the original Cyclone Base Case with $R/L_{T_e} = R/L_{T_i} = 6.96$ where both the ITG and TE modes were excited, a case with $R/L_{T_i} = 6.96$ and $R/L_{T_e} = 0$ where mainly the ITG mode was driven unstable and a case with $R/L_{T_i} = 0$ and $R/L_{T_e} = 6.96$ which was mainly TE dominated. In the linear gyrokinetic analysis it was found that the ITG mode is dominant in the negative R/L_n region for $k_y\rho_s < 0.5$ and that the TE mode was dominant otherwise in the mixed case with background peaking factor of between $PF \approx 3$ at $k_y\rho_s = 0.2$ to 6 at $k_y\rho_s = 1.2$. In a scan in R/L_n and plasma β, β was found to have a stabilizing effect in the mixed and ITG cases in the negative R/L_n region in both GENE and EDWM. Both models indicated a decrease in background peaking factor with increased plasma β in the mixed case. Increasing magnetic shear also had a stabilizing effect in the negative R/L_n region while adding collisions had a negligible effect. In the nonlinear analysis the particle fluxes are negative around $R/L_n = 0.0$ in the electrostatic case. Adding β effects, however, the inward particle flux in the negative R/L_n region decreases in both models while it changes sign to outwards in the positive gradient region, following the linear results which indicated a decrease in background peaking factor. This may have serious consequences for the efficiency of the pellet fuelling scheme in high β plasmas.

Acknowledgments

The simulations were performed on resources provided by the Swedish National Infrastructure for Computing (SNIC) at the PDC Centre for High Performance Computing (PDC-HPC) and on the HELIOS supercomputer system at the Computational Simulation Centre of International Fusion Energy Research Centre (IFERC-CSC), Aomori, Japan, under the Broader Approach collaboration between Euratom and Japan, implemented by Fusion for Energy and JAEA. This work was funded by a grant from The Swedish Research Council (C0338001).

References

[1] Garzotti L, Figueiredo J, Roach C, Valovič M, Dickinson D, Naylor G, Romanelli M, Scannell R and Szepesi G 2014 *Plasma Physics and Controlled Fusion* **56** 035004
[2] Kotschenreuther M, Rewoldt G and Tang W 1995 *Computer Physics Communications* **88** 128–140
[3] Tegnered D, Nordman H, Oberparleiter M, Strand P, Garzotti L, Lupelli I, Roach C, Romanelli M and Valovic M 2016 Gyrokinetic simulations of transport in pellet fuelled discharges at jet *43rd European Physical Society Conference on Plasma Physics, Leuven, Belgium*
[4] Bourdelle C, Garbet X, Imbeaux F, Casati A, Dubuit N, Guirlet R and Parisot T 2007 *Physics of Plasmas (1994-present)* **14** 112501
[5] Baiocchi B, Bourdelle C, Angioni C, Imbeaux F, Loarte A and Maslov M 2015 *Nuclear Fusion* **55** 123001
[6] http://genecodeorg/
[7] Kadomtsev B and Pogutse O 1971 *Nuclear Fusion* **11** 67
[8] Coppi B and Pegoraro F 1977 *Nuclear Fusion* **17** 969
[9] Horton Jr W, Choi D and Tang W 1981 *Physics of Fluids (1958-1988)* **24** 1077–1085
[10] Guzdar P, Chen L, Tang W and Rutherford P 1983 *Physics of Fluids (1958-1988)* **26** 673–677
[11] Romanelli F 1989 *Physics of Fluids B: Plasma Physics (1989-1993)* **1** 1018–1025
[12] Biglari H, Diamond P and Rosenbluth M 1989 *Physics of Fluids B: Plasma Physics (1989-1993)* **1** 109–118
[13] Dannert T and Jenko F 2005 *Physics of Plasmas (1994-present)* **12** 072309
[14] Dimits A, Bateman G, Beer M, Cohen B, Dorland W, Hammett G, Kim C, Kinsey J, Kotschenreuther M, Kritz A *et al.* 2000 *Physics of Plasmas (1994-present)* **7** 969–983
[15] Weiland J 1999 *Collective Modes in Inhomogeneous Plasmas: Kinetic and Advanced Fluid Theory* (CRC Press)
[16] Strand P, Bateman G, Eriksson A, Houlberg W, Kritz A, Nordman H and Weiland J 2004 Comparisons of anomalous and neoclassical contributions to core particle transport in tokamak discharges *31th EPS Conference, London 2004, European Physical Society* vol 28

Journal of Physics: Conference Series **775** (2016) 012014 doi:10.1088/1742-6596/775/1/012014

[17] Merz F 2008 *Gyrokinetic simulation of multimode plasma turbulence* Ph.D. thesis Universitat Munster
[18] Nordman H, Skyman A, Strand P, Giroud C, Jenko F, Merz F, Naulin V, Tala T, Contributors J E *et al.* 2011 *Plasma Physics and Controlled Fusion* **53** 105005
[19] Pueschel M J, Kammerer M and Jenko F 2008 *Physics of Plasmas* **15** 102310
[20] Jenko F and Dorland W 2001 *Plasma physics and controlled fusion* **43** A141
[21] Skyman A, Tegnered D, Nordman H and Strand P 2014 *Physics of Plasmas (1994-present)* **21** 092305

Electron dynamics in the vicinity of Lower Hybrid antennas

L. Valade, E. Caschera, P. Ghendrih, Y. Sarazin and A. Ekedahl
CEA, IRFM, F-13108 Saint-Paul-lez-Durance, France
E-mail: laurent.valade@cea.fr

Abstract. Heating and current drive in tokamaks with Lower Hybrid (LH) waves appears to generate hot spots magnetically connected to LH antennas. Correlated to these spurious patterns of heat deposition, fast electron populations are also reported experimentally. A Vlasov–Poisson (1D, 1V) model is used to address this problem. External drive by waves with appropriate spectrum to describe the physics induced by LH antennas is considered. Fast electron tails, symmetric in velocity, are shown to build-up. In the range considered in this work, the energy content of these tails is found to be proportional to the energy of the excitation LH waves.

1. Introduction

In tokamaks using Lower Hybrid (LH) waves for current drive, hot spots can be observed on components magnetically connected to the LH antennas [1, 2]. These hot spots are believed to be caused by fast electrons accelerated in front of the LH antennas. Indeed, fast electrons have been detected in the Scrape-off-layer (SOL) on magnetic field lines connected to LH antennas by Langmuir probes and soft X-ray measurements on TdeV [3], and by RFA measurements on Tore-Supra [4] and EAST [5]. Modeling indicates however that ions could be accelerated in electric fields that result from charge separation during the electron acceleration process [6], yet without experimental evidence to support this result.

This work presents the first steps towards modeling the interaction of electrons with the k-spectrum of the electric field of an LH antenna. The ultimate goal is to predict the heat flux generated by this process and to find mechanisms for mitigating the effect.

LH antennas are composed of toroidal rows of several tens of wave-guides, stacked in the poloidal direction. If one assumes that the electron density in front of the antenna is uniform in the poloidal direction, each row can be considered identical. The hot spots appear to be produced radially within typically one centimeter [4]. Experimental results are consistent with ballistic transport of accelerated electrons along magnetic field lines until they impinge onto plasma facing components (PFCs) [2, 7, 8]. Therefore, we consider here only the dynamics of electrons along a magnetic field line connected to the LH antenna.

This region of interest can be split into three parts: (i) the front of LH antennas where the edge plasma absorbs a small part – between 1 % and 10 % [3] – of the power, (ii) magnetic field lines where heat flux propagates from LH antennas to PFCs and (iii) strike points where magnetic fields lines intersect PFCs and heat flux is deposited. In this first step, we address the excitation part, which allows one to consider periodic boundary conditions. These are easier to handle

Journal of Physics: Conference Series **775** (2016) 012015 doi:10.1088/1742-6596/775/1/012015

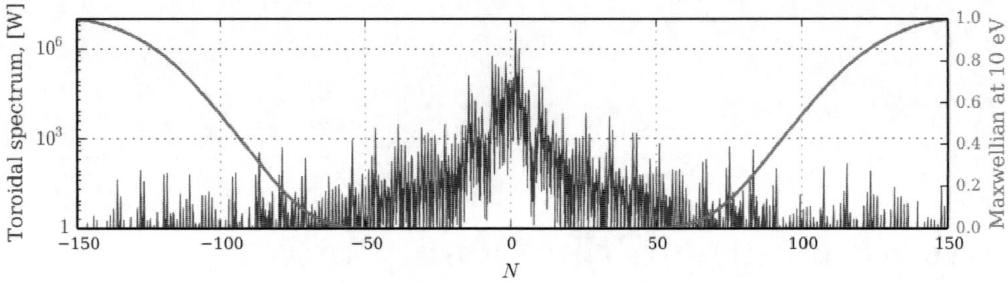

Figure 1. Blue line (LHS scale): spectral density of the LH power. Red line (RHS scale): Maxwellian at 10 eV. Abscissa: toroidal refractive index $N = c/v = ck/\omega$.

numerically. Furthermore, we reduce the simulation domain to two wave guides with opposite phase and given frequency $\omega_{\rm ext}$. The important part of the LH-spectrum in wave vector, figure 1, is then governed by the sharp transition from one wave guide to another, leading to a broadband spectrum at high refraction index (up to typically 100), figure 1. Conversely, the useful part of the spectrum is governed by the synergy between the series of wave guides which yields a refraction index of order 2. While the former part of the spectrum is taken into account by the model, the latter is not. This is consistent with the fact that, in the SOL, particle resonance with the LH waves will only occur in the high refraction index region, figure 1. With the given choice of LH set-up, the simulation domain is rather homogeneous and strong charge separation governed by the expansion of the fast electron cloud will not occur. In agreement with results from the linearized system, it is therefore sufficient to consider frozen ions uniformly distributed at this stage.

The paper is organized as follows, the model and simulation tool are introduced in Section 2, benchmark with respect to the standard Landau damping in Section 3. The excitation by a standing wave is addressed in Section 4 and results for a broad band, wage-guide shaped electric field are presented in Section 5. Discussion and conclusion close the paper, Section 6.

2. Modelling LH wave coupling to SOL plasma

Given the operation range for LH antennas in the SOL plasma, the ratio of antenna size L to collisional mean free path in the parallel direction, $\nu^*_{\rm LH}$, ranges from 10^{-1} to 10^{-3} so that a kinetic model is required to describe the plasma. In a first step, collisional damping of kinetic filamentation is not taken into account, $\nu^*_{\rm LH} \to 0$. One also expects that the electric field varies on scales smaller or comparable to the Debye length so that the Poisson equation is required to determine the electrostatic potential generated by the plasma. Finally, on short time scales, consistent with the high frequency of LH waves, we assume the asymptotic limit $\sqrt{m_{\rm e}/m_{\rm i}} \to 0$ where $m_{\rm e}$ and $m_{\rm i}$ are the electron and ion mass respectively. For this frozen ion background we assume a uniform ion density, which is used to normalize electron and ion densities. Let z be the parallel position normalized by the Debye length $\lambda_{\rm D}$, t time normalized by $1/\omega_{\rm p}$, where $\omega_{\rm p}$ is the plasma frequency, and v the electron velocity normalized by the electron thermal velocity $v_{\rm th} = \sqrt{T_0/m_{\rm e}}$, with T_0 the initial electron thermal energy. The Vlasov–Poisson system for the normalized electric potential, normalized by T_0/e, e being the elementary charge, and normalized electron distribution function f is then:

$$\partial_t f + v\partial_z f + (\partial_z\phi_{\rm p} + \partial_z\phi_{\rm ext})\,\partial_v f = 0, \tag{1a}$$

$$\partial_z^2\phi_{\rm p} = \int_{-\infty}^{+\infty} f\,{\rm d}v - 1, \tag{1b}$$

where $\phi_{\rm ext}$ is the electric potential generated by the LH antenna. Note that the density only appears via the definition of the normalizing scale $\lambda_{\rm D}$ and consequently normalizing frequency $\omega_{\rm p}$. We consider in this paper a periodic system in z of length L and global quasi-neutrality, namely that the electron density averaged in z is equal to 1. To analyze the effect of LH wave coupling to the SOL electron population, it is interesting to determine the evolution of the electron current $j = -nu$, where n and u are respectively the density and mean electron velocity, and total energy $\mathcal{E} = n\left(T + u^2\right)/2$, where T is the electron thermal energy normalized by T_0:

$$\partial_t j - 2\partial_z \mathcal{E} = nE, \qquad \partial_t \mathcal{E} + \partial_z Q = jE. \tag{2}$$

In these balance equations, E is the total electric field, sum of the self-consistent Poisson $E_{\rm p}$ and external $E_{\rm ext}$ contributions, and $Q = \int {\rm d}v(v^3/2)f$ stands for the energy flux. It readily appears that the spatial averages of both j and $\mathcal{E}$, $\langle j\rangle_L$ and $\langle \mathcal{E}\rangle_L$, evolve due to the coupling of the electric field with density and current, respectively:

$$\partial_t \langle j\rangle_L = \langle nE\rangle_L, \qquad \partial_t \langle \mathcal{E}\rangle_L = \langle jE\rangle_L. \tag{3}$$

The balance equation for $\langle n\rangle_L$ is trivially satisfied.

The non-linear dynamics of the driven Vlasov–Poisson system eq. (1) is addressed numerically with the CALVA code (Code for Acceleration of eLectrons in Vicinity of Antennas), which originates from the code developed in [9]. The code adopts the Particle-In-Cell (PIC) method to evolve the distribution function while a second order finite difference scheme is used to solve the Poisson equation. The distribution function f is described by a set of M macro-particles [10] that are coupled via the Poisson equation:

$$f(z,v,t) = \sum_{m=0}^{M-1} M_m S_z\left(z - z_m(t)\right) S_v\left(v - v_m(t)\right), \tag{4}$$

M_m is the assigned (constant) weight of the m-th macro-electron, located in phase space by S_z the shape function in position and S_v the shape function in velocity. The shape functions in position and velocity are chosen to be order one b-splines. The Vlasov equation (1a) can then be replaced by the motion equations of the macro-electrons:

$$\mathrm{d}_t z_m = v_m(t), \qquad \mathrm{d}_t v_m = -E_m(t), \tag{5}$$

$E_m(t)$ is the Lagrangian total electric field, *i.e.* the electric field weighted by the finite-size macro-electron:

$$E_m(t) = \int_0^L \left(-\partial_z \phi(z,t)\right) S_z(z - z_m(t))\,\mathrm{d}z, \qquad \phi = \phi_{\rm p} + \phi_{\rm ext}. \tag{6}$$

3. Simulation of a damped Langmuir wave

In the absence of any excitation source, *i.e.* $\phi_{\rm ext} = 0$, eqs. (1a,1b) reduce to the well-known 2D Vlasov-Poisson system for Langmuir waves. These waves are known to exhibit collisionless Landau damping [11, 12]. Let us linearize the system eqs. (1a,1b) at vanishing exciting field ($\phi_{\rm ext} = 0$) and close to a Maxwellian equilibrium. The equilibrium electric field is null. When expressed in Fourier space $(f,\phi) = \sum_k \int {\rm d}\omega (f_{k,\omega}, \phi_{k,\omega}) \exp[{\rm i}(kz - \omega t)]$, linear solutions are such that angular frequency ω and wave vector k satisfy the following dispersion relation:

$$D(k,\omega) = k^2 + 1 + \zeta\, Z(\zeta) = 0 \tag{7}$$

where $\zeta = \omega/(|k|\sqrt{2})$ and $Z(\zeta) = \frac{1}{\sqrt{\pi}} \int_{\mathbb{R}} \frac{\exp(-\theta^2)}{\theta - \zeta} \, \mathrm{d}\theta$ is the plasma dispersion function [13]. In the case of a periodic system, k is real. Conversely, $\omega = \omega_\mathrm{r} + \mathrm{i}\gamma$ has a non-vanishing imaginary part γ, the Landau damping rate, which accounts for the resonant interaction between waves and particles. It can be computed numerically. This mechanism is suspected to play a dominant role in electron heating by LH waves. In addition, recovering the analytical values of γ constitutes a stringent test for numerical codes. As detailed below, this test is used to benchmark our PIC code.

At $t = 0$, the distribution function is prepared so that the density field is perturbed by a single mode k, $n(z, t = 0) = 1 + \varepsilon \cos(kz)$, leading to the following initial electric potential: $\phi_\mathrm{p}(z, t = 0) = -(\varepsilon/k^2)\cos(kz)$. The small magnitude of the perturbation $\varepsilon \ll 1$ ensures that the system remains in the linear regime. The initial perturbation evolves towards the following damped stationary wave:

$$\phi_\mathrm{p}(z, t) = -\frac{\varepsilon}{k^2} \exp(\gamma t) \cos(kz) \cos(\omega_\mathrm{r} t) \,. \tag{8}$$

Here, $(k, \pm\omega_\mathrm{r} + \mathrm{i}\gamma)$ is the solution of the dispersion relation eq. (7) (both signs of ω_r are solutions, as a result of the symmetry property of the dispersion function, $Z(\zeta^*) = -Z^*(-\zeta)$) with the smallest imaginary part, other solutions with larger values of $|\gamma|$ being damped away on very short time scales.

In figure 2, results of a simulation performed with the CALVA code are compared with theoretical predictions. About $M \sim 26 \cdot 10^6$ macro-electrons were used, with $k = 0.4$ (the domain is $L = 2\pi/k \approx 15.7$ long) and $\varepsilon = 0.005$. $k = 0.4$ corresponds to the high-k / low phase velocity component of LH antennas. Both ω_r and γ are successfully recovered with relative errors of 0.08 % and 0.3 %, respectively. The exponential decay does not reach the floating number precision $\sim 10^{-16}$ in the simulation run. Indeed, it is lower bounded by the intrinsic statistical noise of the PIC method, which decreases as $M^{-1/2}$. In the present case, the exponential decay evolves into a stochastic behavior for $t \gtrsim 50$.

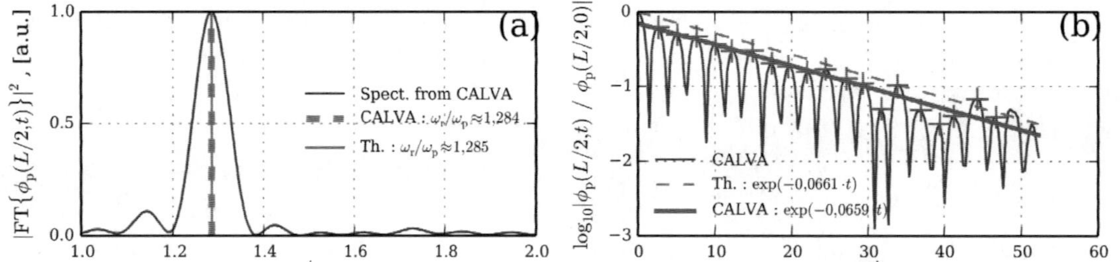

Figure 2. Blue lines are from a CALVA simulation. (a) Fourier transform of the electrical potential $\phi_\mathrm{p}(z = L/2, t)$ (time window: $0 \le t \le 60$). Red line: theoretical prediction for ω_r. (b) Time evolution of $\phi_\mathrm{p}(z = L/2, t)$. Dashed red: theoretical exponential decay.

4. Excitation by a monochromatic plane wave

When forced by an external electric potential ϕ_ext, the system eqs. (1a,1b) exhibits a more complex behaviour: Landau damping now competes with the excitation source. Before addressing the case of a broad excitation spectrum, reminiscent of that of the LH antenna, let us consider the simpler case of an excitation by a monochromatic plane wave $\phi_\mathrm{ext}(z, t) = \phi_{\mathrm{ext},k\omega} \cos(\omega_\mathrm{ext} t - k_\mathrm{ext} z)$.

Provided the source amplitude remains small enough $\phi_{\mathrm{ext},k\omega} \ll 1$, the plasma response can be studied within the linear framework. In this case, one can compute the linear gain $G(k, \omega)$,

Journal of Physics: Conference Series **775** (2016) 012015 doi:10.1088/1742-6596/775/1/012015

defined as the ratio between the amplitude of the response $\phi_{\mathrm{p},k\omega}$ and the amplitude of the excitation $\phi_{\mathrm{ext},k\omega}$. If the initial distribution function is Maxwellian, it reads as follows:

$$G(k,\omega) = \frac{\phi_{\mathrm{p},k\omega}}{\phi_{\mathrm{ext},k\omega}} = -\frac{1+\zeta Z(\zeta)}{k^2+1+\zeta Z(\zeta)} \tag{9}$$

Here, $k = k_{\mathrm{ext}}$ and $\omega = \omega_{\mathrm{ext}}$ (we recall that $\zeta = \omega/(|k|\sqrt{2})$) are the wave vector and angular frequency of the excitation monochromatic wave. The gain G, which is a complex function, is displayed in fig. 3. Both 2D (k,ω) contours and 1D curves are plotted for its magnitude and argument. Green lines correspond to Langmuir wave solutions of eq. (7), namely $(k, \omega_{\mathrm{r}}(k))$. The dashed black line corresponds to the maximum of $|G|$ with respect to ω.
Complexity of the gain function $G(k,\omega)$ stems directly from the balance between real and imaginary parts of the plasma dispersion function $Z(\zeta)$. One can however draw a few properties. In the fluid limit $\omega/k \gg v_{\mathrm{th}}$ (with $v_{\mathrm{th}} = 1$ the normalized electron thermal velocity), the imaginary part of $Z(\zeta)$ tends to be small, so that the denominator of $G(k,\omega)$ will be small for ω solution of the Landau dispersion relation. For small k this yields the Bohm-Gross relation, $\omega_{\mathrm{BG}}^2 = 1+3\,k^2$. This property locates the maximum amplitude of $G(k,\omega)$. In this limit, one can also readily show that the plasma response, *i.e.* the electrostatic potential ϕ_{p}, is in phase with the external potential ϕ_{ext}, while $|G(k,\omega)|$ is proportional to $1/\omega^2$. Conversely, in the adiabatic electron limit $\omega/k \ll v_{\mathrm{th}}$, one finds that ϕ_{p} is opposite to ϕ_{ext} and that $|G(k,\omega)| \approx 1/(1+k^2)$, of order unity for $k \to 0$. The electric field driven by the electrons then cancels out the external electric field.

For a multi-species plasma, the expression of the gain in the linear limit is comparable $G(k,\omega) = -\sum_a g_a(\zeta_a)/\left(k^2+\sum_a g_a(\zeta_a)\right)$, with $g_a(\zeta_a) = c_a Z_a^2(1+\zeta_a Z(\zeta_a))$. Here, the following definitions have been introduced: $c_a = n_a/n_0$, $Z_a = q_a/e$ and $\zeta_a = \sqrt{m_a/m_{\mathrm{e}}}\,\zeta$, with n_a, q_a and m_a the density, charge and mass of particles of species a respectively. In this expression the thermal energies are assumed the same, waving-off this assumption is straightforward. For $\zeta \approx 1$, when electrons are resonant, one thus obtains $\zeta_a \gg 1$. The ions are therefore in the fluid limit with negligible response. This linear analysis result justifies considering the frozen ion limit to address the plasma response to LH drive on short time scales. It also indicates the range of wave vectors required to obtain a large ion response. Indeed with LH, as discussed in the following, ω is fixed while the k-spectrum is governed by the wave-guide shaping. Hence if for the main k_{LH} of the LH launcher one has ω/k_{LH} ranging between 1 and 10, then ion resonance effects will be driven by wave vectors k of order $k_{\mathrm{LH}}\sqrt{m_{\mathrm{i}}/m_{\mathrm{e}}}$, typically 60 times larger, and consequently with a smaller amplitude, at least by a factor $\sqrt{m_{\mathrm{e}}/m_{\mathrm{i}}}$ for a square shaped, hence "perfect", wave guide. Regarding direct acceleration of particles by LH near electric field, it seems therefore reasonable to only consider the electrons.

The spectrum of LH antennas is quasi-monochromatic in frequency and broad in wave-vector. It would therefore make sense to perform a scan in k at fixed excitation frequency ω_{ext}. However, changing the k value while keeping constant the number of grid points per period would require to change the domain size $L = 2\pi/k$ of the simulation. At constant accuracy, the number of macro-electrons should then be modified accordingly. Indeed, in PIC simulations, statistical errors scale like the inverse square-root of the number of macro-electrons M_{cell} per phase–space cell $\Delta v \Delta z$. Given Δv and Δz, the number of macro-electrons M should then scale like $M \propto L$. Scanning the excitation pulsation ω_{ext} instead of the wave vector constitutes an easier (especially less costly in terms of numerical resources) way to benchmark CALVA against linear theory. The scan presented below is performed at $k_{\mathrm{ext}} = 0.4$. At this wave vector, the linear gain $|G|$ is maximum at angular frequency $\omega_G\,(k_{\mathrm{ext}}) \approx 1.285$.
The plasma response ϕ_{p} exhibits a non-monochromatic spectrum, as exemplified in Fig. 4 for

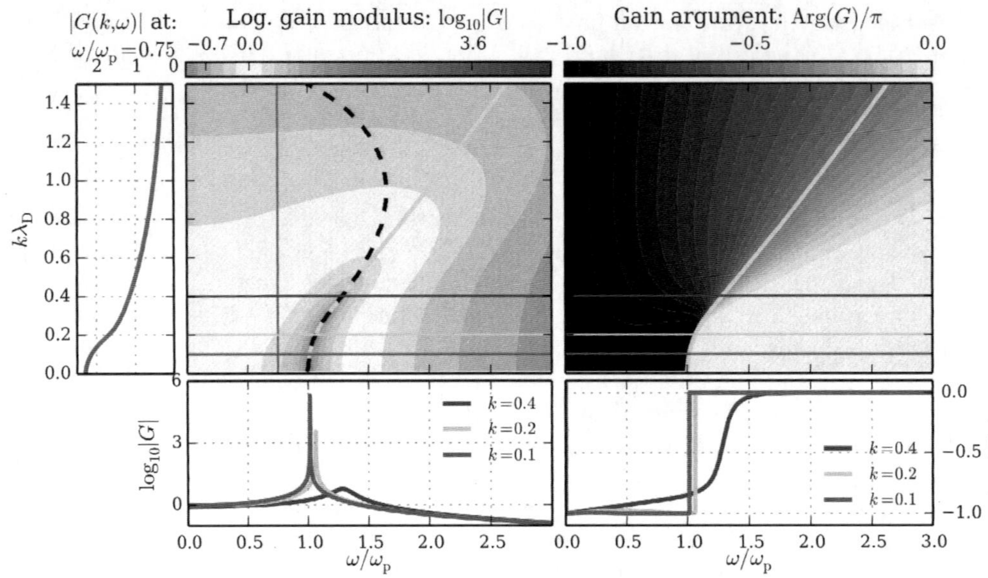

Figure 3. Modulus and argument of the linear gain, eq. (9). ω and k are the pulsation and wave vector of the excitation field ϕ_{ext}.

$\omega_{\mathrm{ext}} \approx 0.964$, which corresponds to $\omega_{\mathrm{ext}}/\omega_G(k_{\mathrm{ext}}) = 0.75$. The Fourier spectrum peak is close to the excitation frequency, but other frequencies are also present. Also, the phase shift (between ϕ_{p} and ϕ_{ext}) is $\approx -0.63\pi$ while $\mathrm{Arg}\left[G(k_{\mathrm{ext}}, \omega_{\mathrm{ext}})\right] \approx -0.86\pi$ is predicted by linear theory. The relative error reaches about 25 %. The discrepancy is of the same order of magnitude for the amplitude of the gain (~ 1.2 in the simulation, as compared to ~ 1.6 from theory).

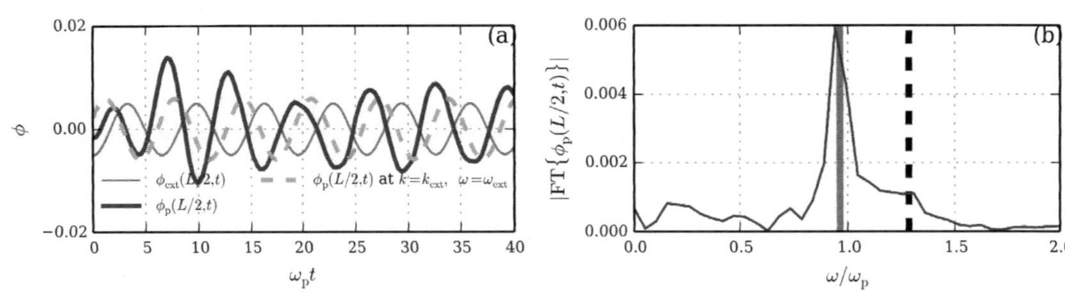

Figure 4. (a) Time evolution at $z = L/2$ of excitation ϕ_{ext} ($\phi_{\mathrm{ext},k\omega} = 0.005$) and response ϕ_{p}, and of the magnitude of the Fourier component of ϕ_{p} at ($k_{\mathrm{ext}} = 0.4, \omega_{\mathrm{ext}} = 0.964$). (b) Fourier spectrum of ϕ_{p} at $z = L/2$. Red line: ω_{ext}. Dashed black: angular frequency at the maximum of $|G|$.

However, it is worth mentioning that the matching between simulations and linear theory predictions is improved when the excitation frequency ω_{ext} departs from the resonant frequency, *i.e.* from the neighborhood of the maximum of $|G|$. This characteristic feature is exemplified in Fig. 5, where the gain is scanned with respect to ω_{ext}, and compared to the linear prediction, eq. (9). Although the magnitude and phase of the gain depart from the linear analytical theory, the resonant frequency matches the analytical one.

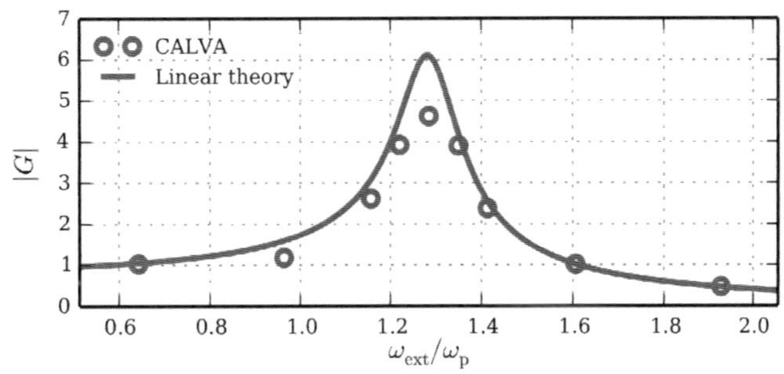

Figure 5. Open circles: gain computed from CALVA simulations for various excitation frequencies ω_{ext} (fixed wave vector $k_{ext} = 0.4$). Plain curve: analytical linear prediction, eq. (9).

5. Excitation by a shaped electric field

The electric field in front of LH antennas is expected to exhibit a broad Fourier spectrum, as exemplified in Fig. 1. In the present work, we consider a simplified expression of the electric field. It captures those features which are essential to explain the acceleration of electrons up to suprathermal velocities.

In this framework, one considers an LH antenna made of 2 wave-guides only, of width $d = 50$ (in Debye units), separated by a zero-width septum and launching the same and uniform power. The width of the periodic simulation domain is thus $L = 2d$. For the sake of simplicity, the phase shift of the electric field from one wave-guide to the other is set at π. This allows one considering a smaller antenna, made of two wave-guides, hence requiring less numerical resources. In this case, the electric field has the following expression:

$$E_{ext}(z,t) = E_0 \cos\left(\omega_{ext} t - \varphi(z)\right), \quad \varphi(z) = \mathrm{int}\left(\frac{2z}{L}\right)\pi, \tag{10}$$

where int(x) stands for the integer part of x. E_{ext} can also be expressed as superimposed monochromatic waves $E_{ext}(z,t) = E_0 \sum_{j\in\mathbb{Z}} \frac{4}{k_j L} \cos\left(\omega_{ext} t - k_j z + \pi/2\right)$, with $k_j = (1+2j)2\pi/L$ the wave vector of wave j. One then finds a series of resonances at resonant velocities $\omega_{ext}/k_j = v_j$ and resonant width $\Delta v_j = 2\omega_{b,j}/k_j$, where $\omega_{b,j}^2 = 4E_j/L$ and $E_j = E_0 \left|1 + G\left(\omega_{ext}, k_j\right)\right|$ takes into account the plasma response via the gain function for each wave-vector. At each resonance a plateau is expected to develop on the width Δv_j with characteristic time $1/\omega_{b,j}$, the trapping time in the island.

With time, electrons can experience large velocity excursions from one resonance location to another, leading to the formation of plateaus of the distribution function in the velocity space. This overlapping effect can lead to merging of the plateaus. As exemplified in Fig. 6(a), the initial Maxwellian then develops heavy tails. There, the plateau extends up to 3 times the thermal speed at $t \approx 200$. This generic mechanism is suspected to be responsible for the acceleration of electrons up to several times the thermal velocity in front of LH antennas. The power transfered from the wave to the electrons, and given by the product of the electron current and the electric field $\langle jE \rangle_L$ eq. (3), scales like the energy injected by the exciting field. This proportionality is evident in Fig. 6(b), where each symbol corresponds to a different simulation at a prescribed magnitude of E_{ext}.

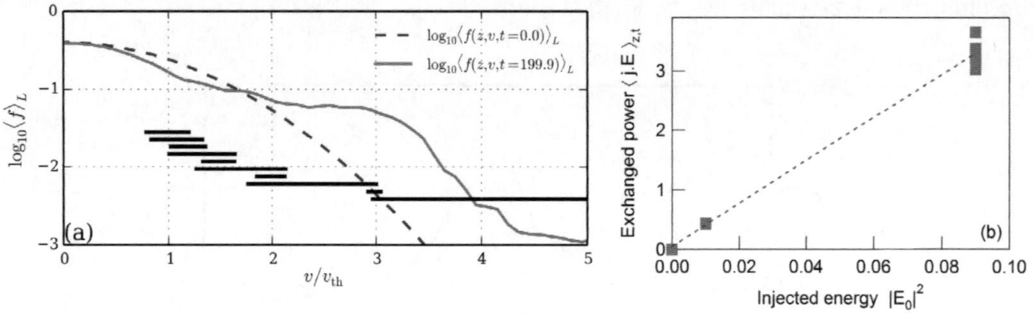

Figure 6. (a) Spatial average of the distribution function at $t = 0$ (dashed blue) and $t \approx 200$ (bold red) in the case $E_0 = 0.3$. Black horizontal lines give the location and width of resonances (see text). (b) Exchanged power between waves and electrons as a function of the injected energy.

6. Discussion and Conclusion

The present analysis of the electron response to the near electric field of Lower Hybrid launchers in the SOL plasma indicates that tails of accelerated electrons can be formed, ranging from one to three times the thermal velocity. In agreement with work published on this issue, these plateaus in velocity space are observed to be driven by an overlap effect of different resonances. The wave vectors and electric field amplitude associated to the latter are governed by the square-shape of the electric field pattern, hence, as expected, by the geometry of the LH launcher. We show here that the kinetic energy of the accelerated electron population is proportional to the LH power. On the short times of the present simulations, the frozen ion assumption is justified. The fast electron population then tends to be trapped by the ions. Acceleration remains possible given the periodic conditions. When addressing the physics that could allow these fast electrons to escape from the launcher vicinity and create hot spots, more demanding self consistent electron-ion simulations are required, including the near antenna region along the field lines. Work along this line is in progress.

[1] M. Goniche, *et al.*, *Nucl. Fusion* **38**, 919 (1998).
[2] P. Jacquet, *et al.*, *Nucl. Fusion* **51**, 103018 (2011).
[3] J. Mailloux, *et al.*, *Journal of Nuclear Materials* **241–243**, 745–749 (1997).
[4] J. Gunn, *et al.*, *J. Nucl. Mater.* **390–391**, 904 (2009).
[5] Y. L. Li, *et al.*, *Phys. Plasmas* **22**, 022510 (2015).
[6] V. Petrzilka, *et al.*, *Czechoslovak Journal of Physics* (Acad. Sci. Czech. Republic, Inst. Phys., Czech Republic, 1999), vol. 49, pp. 127–140. 2nd Europhysics Workshop on the Role of Electric Fields in Plasma Confinement and Exhaust, 19-20 June 1999, Maastricht, Netherlands.
[7] M. Goniche, *et al.*, *Plasma Phys. Contr. Fus.* **46**, 899 (2004).
[8] K. M. Rantamäki, *et al.*, *Plasma Phys. Contr. Fus.* **47**, 1101 (2005).
[9] V. Fuchs, *et al.*, *Nucl. Fusion* **43**, 341 (2003).
[10] G. Lapenta, Particle in cell method, a brief description of the pic method (2009).
[11] L. Landau, *J. Phys. U.S.S.R.* **10**, 25 (1946).
[12] C. Mouhot, C. Villani, *Acta Mathematica* **207**, 29 (2011).
[13] B. D. Fried, S. D. Conte, *The Plasma Dispersion Function. The Hilbert Transform of the Gaussian* (Academic Press, 1961), first edn.

An iterative method to include spatial dispersion for waves in nonuniform plasmas using wavelet decomposition

Pablo Vallejos, Thomas Johnson, Torbjörn Hellsten
Fusion Plasma Physics, EES, KTH, SE-10044 Stockholm, Sweden
E-mail: pablova@kth.se

Abstract. A novel method for solving wave equations with spatial dispersion is presented, suitable for applications to ion cyclotron resonance heating. The method splits the wave operator into a dispersive and a non-dispersive part. The latter can be inverted with e.g. finite element methods. The spatial dispersion is evaluated using a wavelet representation of the dielectric kernel and added by means of iteration. The method has been successfully tested on a low frequency kinetic Alfvén wave with second order Larmor radius effects in a nonuniform plasma slab.

1. Introduction

Plasma waves with perpendicular wavelengths comparable to the ion Larmor radius will experience a non-local response; an acceleration at one point along the gyro orbit will induce a current along the whole orbit. Consequently, the dielectric response is an integral operator, integrating the acceleration along the gyro orbit, which will depend on the wavelength, i.e. it is spatially dispersive.

Numerical methods, such as finite element (FE), finite difference (FD), and Fourier spectral methods, are efficient for solving non-dispersive electromagnetic problems. These methods can be used in certain limits for spatially dispersive problems, e.g. for calculating fast-wave propagation in fusion plasmas during ICRF with negligible spatial dispersive effects [1]. In general, the numerical modelling of waves in spatially dispersive media tends to be significantly more complicated than similar non-dispersive electromagnetic problems, due to its integral character. The inclusion of spatial dispersion effects to all orders in finite Larmor radius was first solved by Sauter et al [2], who derived a set of integro-differential equations for the wave fields in a plane slab which was solved with FE discretisation. Spatial dispersion can also be handled by Fourier spectral methods (see e.g. [3]), but have the disadvantage of producing large and dense matrices. Such matrices are time consuming to invert and memory expensive. Recently, an alternative technique has been proposed for plasma waves with spatial dispersion that uses either FE or FD methods and iterate on the induced current [4].

In this study we extend the method proposed in [5] by generalising the operator splitting. We propose to identify non-dispersive parts by evaluating the dispersive response at an approximate wave vector. The operator splitting is performed between the spatially dispersive and non-dispersive parts of the wave operator. The dispersive part is considered as an inhomogeneous

Journal of Physics: Conference Series **775** (2016) 012016 doi:10.1088/1742-6596/775/1/012016

term in the wave equation, which is solved by means of iteration with Anderson acceleration [6]. The evaluation of the dispersive response is performed using a Morlet wavelet representation.

The paper is organised as follows: In section 2, the iterative procedure is formulated and the relation to kinetic Alfvén waves is described. In section 3, the spatially dispersive response is evaluated using wavelets. In section 4, a numerical example is presented, showing that the iteration procedure works. Conclusions are drawn in section 5.

2. Electromagnetic wave equations with spatial dispersion

The problem we are aiming to solve is a wave equation with a spatially dispersive response

$$\mathcal{L}[\mathbf{E}](\mathbf{r}) = i\omega\mu_0\mathbf{J}_{ant}(\mathbf{r})\,, \tag{1}$$

where

$$\mathcal{L}[\mathbf{E}](\mathbf{r}) \equiv \nabla\times\nabla\times\mathbf{E}(\mathbf{r}) - \frac{\omega^2}{c^2}\mathcal{K}[\mathbf{E}](\mathbf{r})\,, \tag{2}$$

$$\mathcal{K}[\mathbf{E}](\mathbf{r}) \equiv \int\frac{d\mathbf{k}}{(2\pi)^3}\int d\mathbf{r}' K(\mathbf{r},\mathbf{k})\mathbf{E}(\mathbf{r}')\exp\left[i\mathbf{k}\cdot(\mathbf{r}-\mathbf{r}')\right]. \tag{3}$$

The dielectric kernel $K(x,k)$ is described in Ref. [7, 8].

When studying the propagation of a particular wave (e.g. the fast wave during ICRF), a solution to Eq. (1) can be obtained by assuming that the wave-vector is given by an approximate dispersion relation, $k_D(\mathbf{r})$. For example, when modelling ion cyclotron heating a large part of the wave is usually well described by a fast-wave dispersion relation [9]. The wave equation can then be written on a form where the dielectric response is no longer spatially dispersive

$$\mathcal{L}_0[\mathbf{E}](\mathbf{r}) \equiv \nabla\times\nabla\times\mathbf{E}(\mathbf{r}) - \frac{\omega^2}{c^2}K(\mathbf{r},k_D(\mathbf{r}))\mathbf{E}(\mathbf{r}) = i\omega\mu_0\mathbf{J}_{ant}(\mathbf{r})\,. \tag{4}$$

This equation can be solved using standard FE or FD methods [1].

In this report we propose to solve Eq. (1) by first splitting the wave operator $\mathcal{L} = \mathcal{L}_0 + (\mathcal{L} - \mathcal{L}_0)$ and formally rewriting the equation on the form

$$\mathbf{E}(\mathbf{r}) = -\mathcal{L}_0^{-1}\left[(\mathcal{L}-\mathcal{L}_0)[\mathbf{E}](\mathbf{r}) + i\omega\mu_0\mathbf{J}_{ant}\right](\mathbf{r})\,, \tag{5}$$

where $\mathcal{L}_0^{-1}$ can be generated using FE or FD methods. Eq. (5) can be solved using a fixed-point iteration scheme. This formulation is most effective when the spatial dispersion is weak, such that $\mathcal{L}$ and $\mathcal{L}_0$ have similar solutions. However, the formulation is not restricted to this limit. In fact, using a fixed point iteration with Anderson acceleration [6, 4], as used in this report, a wide range of inhomogeneous problems with strong spatial dispersion can be solved.

2.1. Second order ODE describing Kinetic Alfvén waves

To study the properties of the proposed scheme, Eq. (5), we will study the solutions to an ODE of second order. This equation can be derived from Eq. (1) when considering kinetic Alfvén waves in a plasma that is homogeneous along straight field lines, while assuming that the ratio of the ion Larmor radius over the wavelength is small. When aligning the coordinates such that the magnetic field is in the z direction and the perpendicular wave number is in the x direction, the wave equation may be written as [9]

$$n_\parallel^2 E(x) - \int_{-\infty}^{\infty}\frac{dk}{2\pi}\int_{-\infty}^{\infty} dx' \exp\left[ik(x-x')\right] K_{xx}(x,k)E(x') = 0\,, \tag{6}$$

where the K_{xx} is the dielectric tensor component in the (x,x)-direction, $n_{\|}$ is the parallel refractive index and $E(x)$ is the x-directed electric field Fourier transformed in the z-direction. By expanding the dielectric tensor in the perpendicular wave number and neglecting the weak dependence on the parallel wave number, $K_{xx}(x,k) \approx K_0(x) - K_1(x)k_{\perp}^2$, an ODE is obtained

$$\left(\frac{\partial^2}{\partial x^2} + f^2(x)\right) E(x) = 0\,, \tag{7}$$

where $f^2(x) = (K_0(x) - n_{\|}^2)/K_1(x)$.

3. Wavelet representation of the wave equation

Dielectric responses that include finite Larmor radius effects form an integral operator, or alternatively a differential operator of infinite order. To evaluate such operators, the basis for describing the electric field should ideally have inifinite number of derivatives. In inhomogeneous media, a spatially localised basis is preferable for computational efficiency. We therefore propose the use of a Morlet basis and continuous wavelet transform. The Morlet wavelet not only satisfies the conditions above, but also has a narrow Fourier spectrum such that harmonic functions have a narrow wavelet spectra.

3.1. Continuous wavelet transform

The wavelet transform is performed using the basis

$$\psi_{a,b}(x) = \frac{1}{\sqrt{a}}\psi\left(\frac{x-b}{a}\right)\,, \tag{8}$$

where

$$\psi(x) = \pi^{-\frac{1}{4}} e^{-\frac{x^2}{2}} \left(e^{i\sigma x} - \kappa\right) \tag{9}$$

is the complex Morlet wavelet (see Fig. 1) with $\kappa = e^{-\sigma^2/2}$ and $\sigma = 6$. This choice of basis is localised in both real space around $x = b$ and in wave number around $k = \sigma/a$ (the Fourier transform is a Gaussian with width $1/a$, see Fig. 1). The wavelet transform, **WT**, is defined as [10]

$$E_{a,b} = \mathbf{WT}\left[E(x)\right](a,b) = \int_{-\infty}^{\infty} E(x)\psi_{a,b}^*(x)dx\,, \tag{10}$$

with the inverse transform

$$E(x) = \mathbf{WT}^{-1}\left[E_{a,b}\right](x) = \frac{1}{C_\psi}\int_{-\infty}^{\infty}\int_{-\infty}^{\infty} E_{a,b}\psi_{a,b}(x)\frac{dadb}{a^2}\,, \tag{11}$$

$$C_\psi = 2\pi \int_{-\infty}^{\infty} |\hat{\psi}_{a,b}(k)|^2 \frac{dk}{k}\,, \tag{12}$$

where $\hat{\psi}_{a,b}(k)$ is the Fourier transform of $\psi_{a,b}(x)$.

3.2. Wavelet representations of the dielectric kernel

The spatially dielectric response in the wave equation, Eq. (1), can be expressed using a Morlet representation of the electric field (for simplicity, this derivation will be performed assuming an electric field that only depends on a single coordinate x)

$$\mathcal{K}[\mathbf{E}](x) = \frac{1}{C_\psi}\int_{-\infty}^{\infty}\frac{da}{a^2}\int_{-\infty}^{\infty}db\int_{-\infty}^{\infty}\frac{dk}{2\pi}K(x,k)\mathbf{E}_{a,b}\hat{\psi}_{a,b}(k)e^{ikx}\,. \tag{13}$$

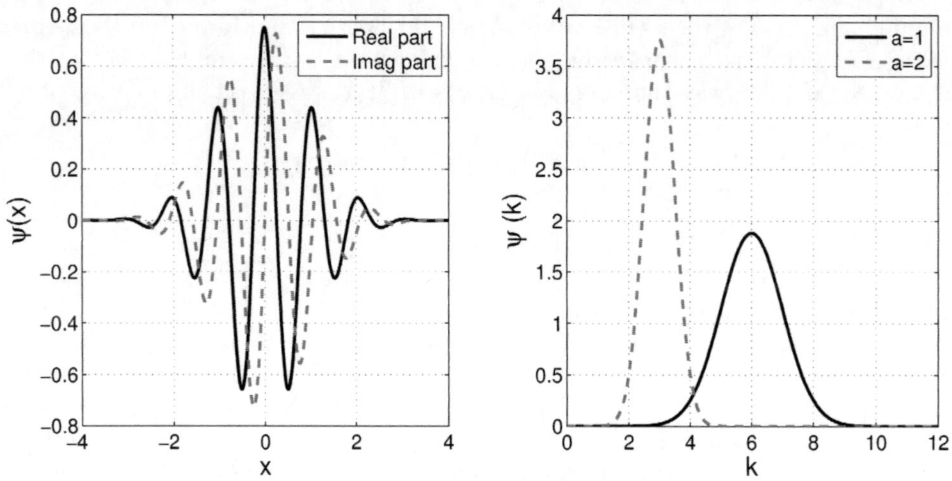

Figure 1. Wavelet basis defined in Eq. (8). *Left*: Solid black and dashed red lines represent the real and imaginary part of $\psi_{1,0}(x)$, respectively. *Right*: Fourier transformed complex Morlet wavelets, $\hat{\psi}_{1,0}(k)$ (solid black) and $\hat{\psi}_{2,0}(k)$ (dashed red).

Since $\hat{\psi}_{a,b}(k)$ is localised around $k \sim \sigma/a$ (the wave number of the eikonal factor in the Morlet basis) one can make the expansion

$$K(x,k) = K\left(x, \frac{\sigma}{a}\right) + \left(k - \frac{\sigma}{a}\right) \frac{\partial K(x,k)}{\partial k}\Big|_{k=\frac{\sigma}{a}} + \frac{1}{2}\left(k - \frac{\sigma}{a}\right)^2 \frac{\partial^2 K(x,k)}{\partial k^2}\Big|_{k=\frac{\sigma}{a}} + \dots \quad (14)$$

Note that when K is a second order polynomial in k, such as for the kinetic Alfvén wave in Eq. (7), this expansion is exact. The dielectric response is then given by an inverse wavelet transform

$$\mathcal{K}[\mathbf{E}](x) = \mathbf{WT}^{-1}\left[K_{a,b}\left(x, \frac{\sigma}{a}\right)\mathbf{E}_{a,b}\right](x) \quad (15)$$

$$K_{a,b}(x,k) = K(x,k) + i\left(\frac{x-b}{a^2}\right)\frac{\partial K(x,k)}{\partial k} + \frac{1}{2}\left(\frac{1}{a^2} - (\frac{x-b}{a^2})^2\right)\frac{\partial^2 K(x,k)}{\partial k^2} + \dots \quad (16)$$

The coefficients in this expansion are Hermite polynomials. In the derivation of Eq. (16), the terms proportional to κ have been neglected, since they give a negligible contribution.

3.3. Wavelet representations of the kinetic Alfvén wave equation

The equation for the kinetic Alfvén wave, Eq. (6), can be expressed using the dielectric response in Eq. (16)

$$f^2(x)E(x) = \mathbf{WT}^{-1}\left[\left\{k^2 + i2\left(\frac{x-b}{a}\right)k - \left(\frac{1}{a^2} - (\frac{x-b}{a^2})^2\right) + \dots\right\}\Big|_{k=\sigma/a} E_{a,b}\right](x). \quad (17)$$

The same equation can be derived by inserting the Morlet representation into Eq. (7)

$$f^2(x)E(x) = -\frac{\partial^2}{\partial x^2}\left\{\frac{1}{C_\psi}\int_{-\infty}^{\infty}\frac{da}{a^2}\int_{-\infty}^{\infty} db E_{a,b}\psi_{a,b}(x)\right\}. \quad (18)$$

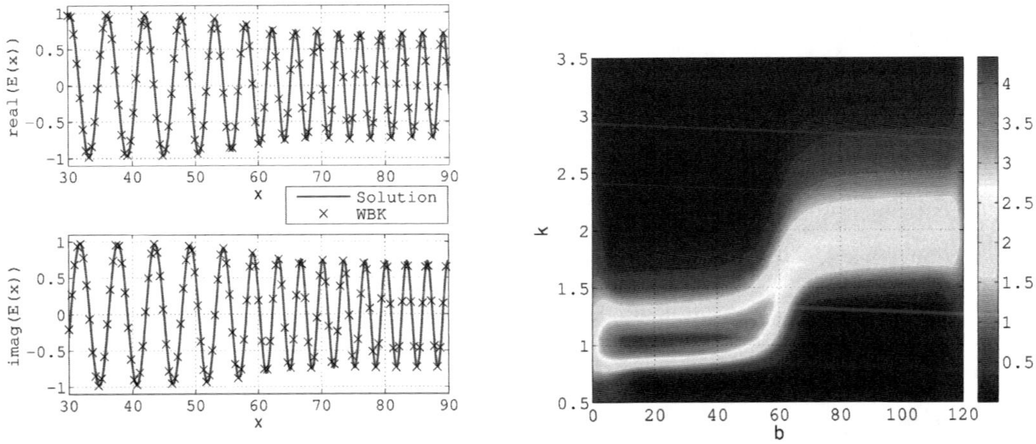

Figure 2. Solution to Eq. (17) and comparisons with the WKB solution of Eq. (18). *Left*: The real and imaginary parts of the electric field in blue, with the WKB solution is indicated by crosses. *Right*: Wavelet representation $E_{a,b}$ of the electric field with $k = \sigma/a$.

4. Results

The ODE in Eq. (17) has been solved on an interval $(0, 120)$ for $f(x) = \frac{3}{2} + \frac{1}{2\pi}\text{atan}([x - 60]/5)$, such that the solution to the local dispersion relation have wavelength between π and 2π.

Morlet wavelets are defined on an infinite interval. Applying them to a finite interval means the wavelet transform is no longer invertable near the boundaries. To ensure that the wavelet transform can be inverted inside our domain, the transform has been performed in an extended domain $(-30, 150)$. In the extended layers, here called the "matched layers", a harmonic solution has been imposed that matches the dispersion relation at the boundary. The matched layers also provides the boundary conditions to the differential equation; in the layer to the left a right propagating wave is imposed with unit amplitude, $E_x = \exp(ikx)$, while at the right boundary a matching procedure is introduced to identify the complex amplitudes of the right and left propagating waves (although in the problems studied the left propagating waves can be neglected).

Numerical solutions of Eq. (17) are illustrated in Fig. 2. The solution in the figure has been compared with a WKB solution ($E(x) \sim \exp(\int^x k(x')dx')/\sqrt{k(x)}$), showing good agreement. More specifically, the solution represents correctly both the amplitude and the phase of the wave in an inhomogeneous domain. The wavelet spectrum, $E_{a,b}$ in Fig. 2, illustrates how the wavelet representation is localised near wavenumbers satisfying the dispersion relation at $x = b$. The spectrum is calculated on the finite domain $(0, 120)$, excluding the matched layers, causing pollution (a numerical widening of the spectrum) in the wavelet-spectrum near the boundaries $b = 0$ and $b = 120$. This illustrates the importance of the matched layers to provide a clean transform.

While Eq. (13) provides an exact response, it is computationally more expensive than the expanded formulation in Eq. (16). To understand the type of error generated by expanding the dielectric response in Eq. (14), we have compared solutions with different order expansions. The results are shown in Fig. 3. In figure a) and b), both the second and third terms in Eq. (16) are neglected (thus approximating $K(x, k) = K(x, \sigma/a)$) and the solutions exhibit strong oscillations and a non-negligable offset in both the frequency and amplitude. In figure c) and d), the first order term in $(k - \sigma/a)$ has been added (second term in Eq. (16)), which reduced both the

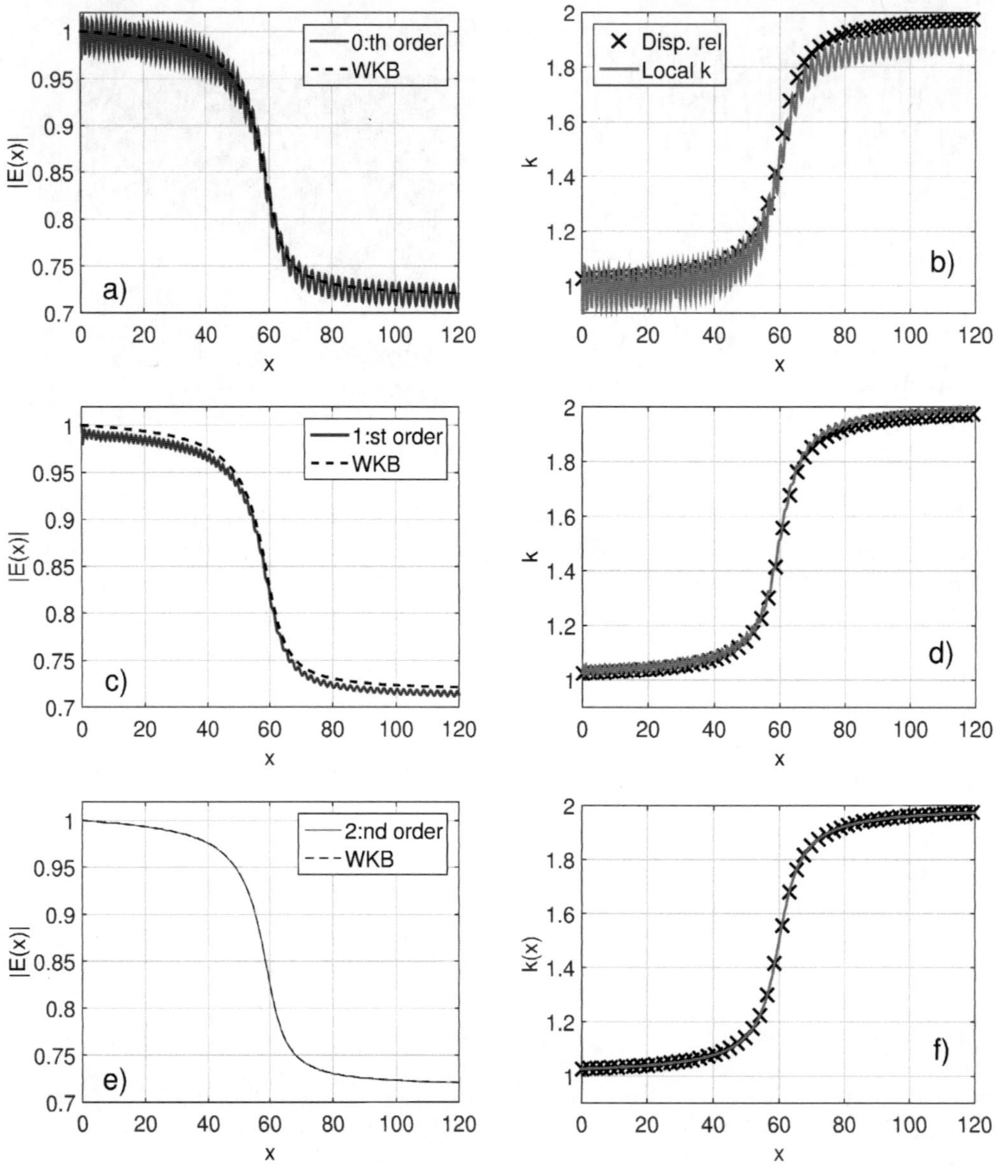

Figure 3. Solutions of Eq. (17) represented in terms of the wave amplitude, $|E(x)|$, and the wave vector, $k(x) = \partial_x Im[\ln(E(x))]$. Figure a) and b) are evaluated when neglecting terms of order σ and σ^2. Figure c) and d) are evaluated when neglecting terms of order σ. Figure c) and d) are evaluated with all terms.

oscillations and the offset. Finally, in figures e) and f), all terms in Eq. (16) are kept, yielding good agreement with WKB solution.

5. Discussion and conclusions

A novel iterative technique for solving the spatially dispersive wave equation, Eq. (5), has been proposed. The technique has the potential of including spatial dispersive effects in a simple manner. Using fixed-point Picard iterations, this equation tends to be unstable. However,

Journal of Physics: Conference Series **775** (2016) 012016 doi:10.1088/1742-6596/775/1/012016

solutions can be found using Anderson acceleration. For Picard-unstable problems a large number of iterations is required to find a solution using Anderson acceleration; e.g. the solutions presented in section 4 were found after about 100 iterations. Initial studies indicate that the number of iterations depend mainly on the complexity of the solution, while the dependence on the initial guess and the grid resolution is weak. The present method is rather slow, however, there are several possibilities for optimization.

The numerical complexity of the wavelet representation can be simplified by expanding the dielectric response function for wave numbers near the fundamental wave number of the Morlet wavelet, σ/a. For dielectric responses of finite order in k, the expansion can be made to exactly represent the operator. However, for response tensors with all-order FLR effect a truncated expansion is of interest. We have shown that truncations neglecting second order may cause an oscillation of the wave amplitude, while keeping only zeroth order terms may give rise to an error in both the wave number and the amplitude of the solution. The higher order terms in this expansion describe the spectral width of Morlet wavelet. The nature of these oscillations are still being investigated, however, our conclusion is that the spectral width of the Morlet wavelet has to be taken into account to obtain a converged solution.

The Morlet wavelet has several attractive features, such as differentiability, localisation in space and wave number. However, the Morlet wavelet basis has redundancy in the representation of continuous functions, such that there is more than one way to represent the same signal. While it is still possible to generate an inverse transform, the wavelet representation tends to be computationally inefficient. The wavelet representation redundancy can be reduced by smart reduction of the grid parameters in wavelet space, i.e. $\{a, b\}$, when performing the inverse transform.

References

[1] Villard L, Appert K, Gruber R and Vaclavik J 1986 *Computer Physics Reports* **4** 95
[2] Sauter O and Vaclavik J 1994 *Computer Physics Communications* **84** 226
[3] Jaeger E F, Berry L A, D'Azevedo E, Batchelor D B and Carter M D 2001 *Physics of Plasmas* **8** 1573
[4] Green D L and Berry L A 2014 *Comput. Phys. Commun.* **185** 736
[5] Hellsten T, Johnson T and Vallejos P 2014 *Journal of Physics: Conference Series* **561** 012010
[6] Walker H F and Ni P 2011 *SIAM J Numer. Anal.* **49** 1715
[7] Smithe D 1989 *Plasma Phys. Controlled Fusion* **31** 1105
[8] Dumont R J, Phillips C K and Smithe D N 2005 *Phys. of Plasmas* **12** 042508
[9] Stix T H 1992 *Waves in Plasmas* (Springer-Verlag New York)
[10] Daubechies I 1992 *Ten lectures on wavelets* (Philadelphia, PA, USA: Society for Industial and Applied Mathematics)

Institute of Physics
Dirac House, Temple Back
Bristol BS1 6BE UK

ISSN: 1742-6588
ISBN 978-1-5108-3388-3

9 781510 833883

CURRAN ASSOCIATES INC.
proceedings.com